Maya Salt Production and Trade

Maya Salt Production and Trade

Anthony P. Andrews

The University of Arizona Press
Tucson, Arizona

About the Author

ANTHONY P. ANDREWS spent two years in the Maya area of Mexico and Central America under a grant from the National Geographic Society surveying Maya salt sources while working on his master's and doctoral degrees at the University of Arizona in Tucson, Arizona. His undergraduate studies, which included field work in South America, were completed at Harvard University. Dr. Andrews grew up in Yucatán, Mexico, where he has been engaged in archaeological research since 1971. He has coauthored two monographs and written numerous articles as a result of his research. Dr. Andrews became Assistant Professor of Anthropology at New College of the University of South Florida in 1981.

THE UNIVERSITY OF ARIZONA PRESS

This book was set in 10/11 V-I-P Times Roman
Manufactured in the U.S.A.

Library of Congress Cataloging in Publication Data

Andrews, Anthony P.
Maya salt production and trade.

Bibliography: p.
Includes index.
1. Mayas—Salt. 2. Indians of Mexico—Salt.
3. Indians of Central America—Salt. I. Title.
F1435.3.S24A52 1983 338.2′7632′0972 83-9306

ISBN 0-8165-0813-5

to Barbara

Contents

A Word from the Author	xi
1. Salt and the Maya	1
History of Maya Salt Research	3
Mayan Uses of Salt	8
Sources of Salt	16
2. The Salinas of Yucatán	22
The Northern Salt Beds	22
Minor Sources	41
The Salinas of the Gulf Coast	47
3. The Salinas of Chiapas	51
The Highlands	51
The Pacific Coast	62
4. The Salinas of Guatemala	68
The Pacific Coast	68
The Highlands	77
The Lowlands: Salinas de los Nueve Cerros	95
5. The Salinas of El Salvador	102
6. The Evolution of Salt-making Technology	108
Sal Solar	108
Sal Cocida	110
Seasonality and Specialization	112
7. History of Production and Trade	114
Production	114
Prehispanic Trading Spheres	116
Development of the Salt Trade Networks	118

Formative Period 119
Classic Period 123
Terminal Classic and Postclassic Periods 125
Post-Conquest Native Salt Trade 130

8. Salt, Economics, and Maya Civilization 132
Trade and Rise of Maya State 133

Appendix A: Early Seventeenth-Century Salt Production in Yucatán 136

Bibliography 139

Index 167

FIGURES

1.1 Map of the Maya Area 2
1.2 Distribution of Salt Sources in the Maya Area 17
2.1 Map of Northern Yucatán 22
2.2 Aerial View of the Salinas of Xtampú, Yucatán 24
2.3 Salt Pans at Chavihau, Yucatán 24
2.4 Harvested Salt at San Crisanto, Yucatán 25
2.5 Map of the Celestún Peninsula 26
2.6 Map of Northwest Coast of Yucatán 27
2.7 Map of Central North Coast of Yucatán 27
2.8 Map of Northeast Coast of Yucatán 28
2.9 Aerial View of Las Coloradas, Yucatán 29
2.10 Salt Production of Yucatán, A.D. 1600–1980 40
2.11 Map of Northeast Corner of the Yucatán Peninsula 43
2.12 Distribution of Salt Sources in Gulf Coast Region 48
3.1 Map of Chiapas 52
3.2 Salinas de Portatenco, La Concordia, Chiapas: Solar Evaporation Pans 53
3.3 Salinas de Portatenco, La Concordia, Chiapas: Solar Evaporation Pans With Stick Frames for Star of David Artifacts 54
3.4 Salinas de Portatenco, La Concordia, Chiapas: Star of David Made of Salt 55
3.5 Map of Ixtapa – San Cristóbal Region, Chiapas 57
3.6 Benequenes, Salt Loaves from Ixtapa, Chiapas 58

3.7	Salinas de Punta Flor, Arriaga, Chiapas: Tapescos and Storage Tank, or Taza	64
3.8	Salinas de Punta Flor, Arriaga, Chiapas: Pilas, Solar Evaporation Pans With Recent Salt Harvest	65
4.1	Map of Guatemala	69
4.2	A Perol, Iron Vessel Used for Cooking Brine	70
4.3	Aerial View of Salinas San José, Puerto San José, Guatemala	72
4.4	Map of Lake Amatitlán, Guatemala	78
4.5	Map of Northwestern Guatemala	80
4.6	San Mateo Ixtatán, Huehuetenango: View of Salinas	81
4.7	San Mateo Ixtatán, Huehuetenango: Women Transporting Brine	83
4.8	San Mateo Ixtatán, Huehuetenango: Tinajas, Water Jars	83
4.9	San Mateo Ixtatán, Huehuetenango: An Apaste	84
4.10	San Mateo Ixtatán, Huehuetenango: A Pilón of Salt	84
4.11	Sacapulas, Quiché: Cajón Used for Leaching Playa Soils	90
4.12	Map of Salinas de los Nueve Cerros Region	96
4.13	Salinas de los Nueve Cerros, Alta Verapaz, Guatemala: Classic Period Brine Storage Vessels	97
5.1	Map of El Salvador	103
5.2	Salinas Ayacachapa, Acajutla Region, El Salvador: Iron Perol	104
5.3	Salinas Ayacachapa, Acajutla Region, El Salvador: Abandoned Asoleadcros	105
5.4	Aerial View of Salinas El Aceituno, Jiquilisco Estuary, El Salvador	106
7.1	Salt Trading Spheres of Maya Area and Major Salt Trade Routes of the Classic Period	117

TABLES

1.1	General Chronology of Maya Area	6
7.1	Prehispanic and Modern Salt Production in the Maya Area	115
7.2	Chronology of Exploitation of Salt Sources in the Maya Area and Gulf Coast Region	120
A.1	Early Seventeenth-Century Salt Production in Yucatán	137

A Word From the Author

Mankind can live without gold, but not without salt

Cassiodorus, Fifth Century A.D.

SCHOLARS INTERESTED IN THE ROLE OF TRADE in Maya civilization have long recognized that salt was one of the most important items of exchange in the prehispanic Maya trade networks. Since 1970, a pronounced increase in studies on Maya trade has occurred, most of which stress the critical role this resource has played in the events which shaped Maya history. Given this high level of interest, it was with some surprise that I discovered that information regarding salt sources, production, and trade was fragmentary at best. Even more surprising was the dearth of quantitative data: little or no information was available on the amounts of salt produced in different regions. Thus, while widely proclaiming the importance of salt as a trade resource, scholars, in fact, had no concept of the magnitude of the salt trade.

This book addresses these problems. In 1974 I began gathering information on the distribution, production, and trade of salt in the Maya area. Often, the only record of a given source was a brief mention in an obscure travel account; in many localities, information was lacking on salt production methods, quantities produced, or the length of time the source had been exploited. Since the 1940s, large coastal salt producers using modern solar evaporation techniques have gradually taken over the salt markets and edged out smaller competitors employing traditional, primitive salt-working methods. Native salt-making industries that have functioned for centuries are quickly disappearing throughout the Maya area. Anxious to glean what information was still available, I conducted a field survey of salt sources from May 1976 to May 1978.

In the course of the survey, I traveled more than 100,000 kilometers. The area of study encompassed past and present salt-making localities along the Gulf and Caribbean coasts of Mexico and Belize, and those located in the highlands, lowlands, and along the Pacific coasts of Chiapas, Guatemala, and El Salvador. In addition to locating each source, I interviewed hundreds of informants, including government officials, salt company executives, owners of salt works, and countless saltworkers, traders, and retailers. Particularly valuable were interviews with elderly people, often the sole source of information in places where salt works had long been abandoned.

Historical and archaeological research were a concomitant part of the survey, and I conducted an extensive search for materials in libraries and archives in Yucatán, Chiapas, Guatemala, and El Salvador. I also consulted with numerous historians and archaeologists, and where time and circumstances permitted, made limited archaeological surveys to gather evidence of the prehispanic exploitation of individual sources. The survey's aims were to locate and describe each salt source and techniques of exploitation. I also attempted to obtain the history of the production of each source, including history of the trade of each source; data on salt consumption in the Maya area were also gathered.

Documentation of all aspects of any given salt source was not always possible. Still, the survey produced a large body of data, which, while far from being exhaustive, permits a much broader reconstruction of the salt industry and trade than previously possible.

ACKNOWLEDGMENTS

Several institutions made this project possible. Foremost was the National Geographic Society, whose financial support (Grant No. 1555) enabled me to conduct a survey of Maya salt sources. Further cooperation and assistance came from several government institutions: the Centro Regional del Sureste of the Instituto Nacional de Antropología e Historia in Mexico, the Instituto de Antropología a Historia in Guatemala, and the Administración del Patrimonio Cultural in El Salvador. I am also indebted to the staffs of several Middle American libraries, in particular the Menéndez Library in Mérida, the Casa Na-Bolom Library in San Cristóbal de las Casas, the Archivo General de Centro America in Guatemala City, and the Dirección General de Estadística in San Salvador.

The study would not have been possible without the assistance of numerous individuals whose hospitality, cooperation, and enthusiasm were largely responsible for the project's success. Since several hundred people aided me in one way or another, it is, unfortunately, impossible to acknowledge them appropriately: from saltmaker to government official, their contributions are important.

To William L. Rathje I owe special acknowledgment, for it was his work, in great part, that inspired this project. He and several other colleagues at the University of Arizona provided advice, inspiration, and encouragement in the course of the research: T. Patrick Culbert, Richard A. Thompson, Jane H. Underwood, and Robert McC. Netting.

In the field, assistance came from all quarters. My greatest debt is to Joann Andrews for providing me with a base of operations in Mérida and unflagging support throughout the project. Many other people helped my work in Yucatán: Norberto González C., director of the Centro Regional del Sureste, and Silvia Garza T. de González provided the facilities for conducting extensive surveys along the coasts of the peninsula, as well as personal and professional help in countless ways. I am similarly indebted to Joseph W. Ball, Alfredo Barrera Rubio, Antonio and Margarita Benavides, James Callaghan, Tomás Gallareta, David A. Gilder, David Kelley, Ed Kurjack, Alfonso Llanes, Luis Millet, José Palomeque, Agustín Peña, Luis Ramírez, Fernando and Reta Robles, Joaquín Roche, Peter Schmidt, Eduardo Toro, and David Vlcek.

In Chiapas, the staff of the New World Archaeological Foundation offered hospitality and assistance on a variety of different matters: my thanks go to Gareth Lowe, Thomas Lee, Eduardo Martínez, Susana Ekholm, Barbara Voorhies, and Kristi Butterwick. Additional data on the Chiapanec salt industry came from Donald McVicker, Jim Nations, Carlos Navarette, Maricruz Paillés, Robert Wasserstrom, and Evon Z. Vogt. Gertrude Blom in San Cristóbal de las Casas kindly provided me with photographs (taken by Frans Blom in 1945) of the now-extinct salt works of La Concordia.

In Guatemala, official cooperation came through the kind offices of Luis Luján Muñoz, of the Instituto de Antropología e Historia; furthermore, I am greatly indebted to Rafael and Vivian Morales, Ed and Jenny Shook, Manuel Rubio Sánchez, Francis Polo Sifontes, Sam Bonis, Felix Móntes Cosio, Horacio Alejos, Alain Ichon, Robert Carmack, Nick Hellmuth, and Julio Beltranena for their assistance in a variety of ways in the course of my work in Guatemala.

The highpoints of my travels in Guatemala were my visits to Salinas de los Nueve Cerros, a remote prehispanic salt-working site located deep in the jungles of the Alta Verapaz. This site was being excavated by Brian Dillon in 1977 and 1978, and his hospitality made the visits to his camp at Tortugas very enjoyable. Working with Brian on matters pertaining to salt over the last few years has been a pleasure. I would also like to thank him for his permission to reproduce the photograph in Figure 4.13.

In El Salvador, I would not have accomplished much without the guidance and hospitality of Stanley and Tita Boggs. Roberto Huezo, of the Administración del Patrimonio Cultural, also offered assistance, as did Jacobo Handall, of Usulutan, who explained the workings of the past and present salt industry to me.

Almost every field archaeologist in the Maya area owes a debt to Hal and Alberta Ball and their wonderful flying machine, the *Quetzal*. This project was no exception: many of the aerial photographs of salt-making localities in Yucatán, Guatemala, and El Salvador were taken from their plane.

Finally, many colleagues have offered advice and brought relevant data to my attention. I would like to thank Will and Patricia Andrews, Ursula Ewald, David Freidel, Robin Robertson, Robert Hill, Grant Jones, Arden King, David Phillips, Merle Greene Robertson, Robert Sharer, Raymond Sidrys, Ed Sisson, Barbara Stark, Matthias Strecker, Rick Wilk, and Robert Wauchope.

This book evolved from my 1980 dissertation. From the first page of that work to the final page of this book, I have benefitted in countless ways from the assistance of my wife, Barbara, who, in addition to her aid in the research, edited the entire text, prepared all the illustrations, and proofread the final copy.

My final debt is to Rose Houk, Marshall Townsend, Elizabeth Shaw, and other members of the staff of the University of Arizona Press, whose patient advice and assistance saw this book to its final form.

ANTHONY P. ANDREWS

ABBREVIATIONS LIST

The following citation abbreviations appear in the text:

AGCA	Archivo General de Centro America, Guatemala
AGI	Archivo General de Indias, Seville
DGEYC, *AE*	Dirección General de Estadísticas y Censos, *Anuarios Estadísticos,* San Salvador
RG, DGE	República de Guatemala, Dirección General de Estadística
RG, DGE, *M*	República de Guatemala, Dirección General de Estadística, *Memorias*
RG, MF, *M*	República de Guatemala, Ministerio de Fomento, *Memorias*
RG, MHYCP, *M*	República de Guatemala, Ministerio de Hacienda y Crédito Público, *Memorias*
RG, SF	República de Guatemala, Secretaría de Fomento
RY	Relaciones de Yucatán

1. Salt and the Maya

SALT HAS ALWAYS PLAYED A CURIOUS ROLE in the history of human affairs. It requires relatively little energy to expoit and is easily transported. Its availability at low cost in most markets of the world has led most societies to take it for granted. Paradoxically, salt, or sodium chloride, is vital to the human body and its nutrition: an imbalance or absence of sodium can lead to death. Moreover, scarcity of salt can threaten not only the individual, but the life of a community or nation. Sudden shifts in supply and demand have created havoc within the trading networks of empires, and competition over salt has led to war.

Many societies have satisfied their need for sodium through indirect dietary means. The Eskimo, in whose diet sodium-rich meat and marine resources predominate, has little need for mineral salt. Such is often the case among hunting groups. The Siriono tribe of Bolivia, for example, had a "salt-free" diet prior to their contact with western society (Holmberg 1969:84–89). In modern western societies prepackaged foods, meat, and fish can satisfy basic sodium needs.

Plants, however, have a very low sodium content. Thus, inland agricultural populations who subsist on vegetables and cereals need to supplement their meals with mineral salt. When they live in tropical zones and are involved in hard labor, their need for salt is much higher. This is the case with the Maya.

In the Maya area of Middle America (fig.1.1) salt has been an important trade item for more than two thousand years. In fact, the Maya salt trade played a prominent role in the development and demise of Classic Maya civilization, and the rise of states of the Postclassic period that preceded the arrival of Columbus in the New World. The reason for this is straightforward: some regions had salt sources while others did not.

While nearly every Maya scholar who has written about prehispanic commerce has commented on salt's importance in the trade networks, little in-depth knowledge of this resource is available. Because it is not "visible" in the archaeological record, salt is a resource that can only be studied indirectly. Indeed, most of our knowledge of the use of salt in prehispanic times derives from ethnohistoric accounts and records. Although archaeologists can identify salt-making sources from certain unique features and artifacts,

Fig. 1.1. The Maya area

they cannot trace the trade routes of mineral salt. No specialized containers for transporting salt can be found: the baskets, sacks, and dugout canoes used are perishable items. Nor did the Maya have saltshakers and related items that might have signaled salt's presence. Unlike other advanced ancient cultures such as the Mesopotamians or Aztecs, the Maya left no trace of their commercial records. Aztec tribute lists prominently display a salt symbol, but as yet no "salt glyph" has been identified in the Maya's sophisticated hieroglyphic script.

When the Spanish arived in Yucatán in the sixteenth century, they encountered a large-scale salt industry and far-flung trade of this resource. Its uses permeated daily life—as a dietary item, an article of barter, and an accessory to ritual and medical practices. Thousands were involved in the exploitation of salt sources, and its trade and transportation to markets several hundred miles away required countless bearers and boatmen. These efforts were coordinated by legions of merchants both small and powerful, and the product reached retailers in the most remote villages.

Diverse groups often fought to control profitable salt sources and their associated trade networks. On the plains of northern Yucatán, in the highland valleys of Chiapas and Guatemala, and in the jungles of the southern Péten, competition over salt frequently led to wars and changed political destinies. Repercussions from such events, which disrupted the long-distance trade networks, were felt in households throughout the land.

During the Colonial period, salt continued to be of major economic importance. Royal questionnaires preceded several schemes to tax the industry, and throughout the conquered lands salt, along with maize, cotton, and cacao, became a standard tribute item.

Natives and Spaniards constantly vied for control of salt, and archives are replete with resulting litigation. The Spanish tended to turn a deaf ear to native protests of not-infrequent infringement of comunally owned salt sources. In Yucatán, for example, Spaniards monopolized some of the richest north coast beds whose salt they then traded to Cuba and central Mexico. As use of salt in the silver-refining process created new demand, Yucatecan sources were further exploited throughout the Colonial period.

The importance of this industry has attracted the attention of many scholars who have significantly contributed to our understanding of the role of salt in the development of Maya culture.

HISTORY OF MAYA SALT RESEARCH

Pioneer Studies and Industrial Surveys

The importance of salt in prehispanic Mesoamerican trade networks was recognized by many early colonial chroniclers (cf. Shahagún 1975; Landa, in Tozzer 1941), a fact noted in the earliest papers on

Mesoamerican trade (Bastow 1897; Carrillo y Ancona 1897). However, the first major study of the role of salt in prehispanic Mesoamerica is that of Miguel Othón de Mendizábal, which appeared in 1929 (a second edition appeared in 1946; for summaries see Mendizábal 1930, 1946b). This study includes a broad survey of *salinas* throughout Mexico and a compilation of ethnohistoric data on each salt source.

Mendizábal believed that the geographical distribution of early Mexican populations was largely determined by the distribution of salt sources. He also suggested that control of salt sources was an essential prerequisite to the evolution of complex societies. He also believed that many early population migrations were caused by the need to find adequate supplies of salt. While Mendizábal's interpretations are somewhat simplistic and unabashedly deterministic, his massive compilation of ethnohistoric data has proven invaluable to researchers. For this work alone he deserves the title of "Father of Mesoamerican Salt Studies."

Unfortunately, Mendizábal's survey treats the Maya area in a peripheral manner. Guatemala, Belize, and El Salvador were not covered, and the data on Yucatán and Chiapas are fragmentary.

The earliest studies of Maya salt production and trade are those of Frans Blom (1932), Ralph L. Roys (1943), and Felix McBryde (1947). The works of Blom and Roys, like that of Mendizábal, draw primarily on ethnohistoric sources and deal mainly with the salt beds of northern Yucatán, the largest in Mesoamerica. Blom and Roys list the major sources and outline the primary long-distance salt trade routes in use between northern Yucatán and areas to the south at the time of conquest.

McBryde's 1947 study of southwest Guatemala, which is far more specific, deals with a more limited region and employs both historic and ethnographic data. He describes salt-making technology along the Pacific coast (Tahuexco) and in the highlands (Sacapulas) and includes data on the trade and marketing of the salt from these sources. Aside from these works, scant detailed information was published on salt prior to 1950. Several industrial surveys of the salt sources of Mexico and Central America have been conducted, however.

The first industrial survey of Mexican salt sources, compiled by José Zarate, appeared in 1917 under the title *Las Salinas de México y la Industria de la Sal Común.* This survey, which included an extensive description of the salinas and detailed discussions of salt-working technology was, for three decades, the major reference source on the Mexican salt industry; in the course of compiling his own historical survey, Mendizábal relied heavily on this work.

In 1946 Raúl Lozano García published a new survey of the Mexican salt industry, *Estudio Tecnológico de la Industria de la Sal en México.* This work included a discussion of modern innovations in salt-making technology, and has served since as a general industry manual.

At the regional level, surveys of salt sources appear only as part of a few local geographic or economic studies. Of particular interest is a voluminous study of the economy of Chiapas by Moisés de la Peña (1951) which includes a detailed description of the now moribund salt industry of that state.

In Central America, industrial data are relatively scarce. A commercial directory of Guatemala, compiled by José Quiñonez (1930) lists twenty-nine sources of mineral salt in the highlands. After having spent considerable time attempting to locate each source, I concluded that his information is not reliable. Many of the salinas he lists do not —and never did—exist. Another source on the Guatemalan industry is the annual reports of the Consorcio Salinero de Guatemala, an association of coastal saltmakers who controlled the national industry before World War II. These annual reports for 1942 to 1945 record the production of the Pacific coast salt works during that period (Consorcio Salinero de Guatemala 1942–1945). Finally, unpublished statistics on the salt production of Guatemala from 1950 to 1972 are kept at the Dirección General de Estadística in Guatemala City (RG, DGE, n.d.).

In El Salvador, the only reliable industrial records of the salt industry are found in the *Anuarios Estadísticos* of the Dirección General de Estadística y Censos (DGEYC, *AE*), which contain annual statistics on national salt production from 1944 to the present.

Finally, a more recent, albeit incomplete, listing of Middle American salt sources can be found in Lefond's *Handbook of World Salt Resources* (1969).

Research in the 1960s and 1970s

Following the pioneering works of Mendizábal, Blom, Roys, and McBryde, many years passed before any further research was conducted on the Maya salt industry. During the 1950s only one major contribution on salt appeared, namely Roy's study of the political geography of the Yucatán Maya. This work included data on the management, jurisdiction, production, and related tribute system of the north coast salt works at the time of the Spanish conquest.

Several years later Sir Eric Thompson (1964) published an important study on Maya trade, the first major synthesis of the subject since the earlier works of Blom and Roys. Thompson discussed the salt trade at length and summarized most of the data then available on the topic. A revised edition of this work, which appeared in 1970, is still the best general introduction to the subject of Maya trade.

Another synthesis of Mesoamerican trade data prepared by Brian Dillon (1975) focuses on perishable goods and reviews data on salt; it serves as an informative supplement to Thompson's study.

Several field projects of the 1960s and 1970s have shed light on prehispanic salt sources and saltmaking in different regions of the Maya area. A few prehispanic salt-making sites have been identified in Honduras (Baudez 1973) and Costa Rica (Coe 1962), but the oldest ones are located in Guatemala and Yucatán. In the course of their survey and excavations in the Ocós region on the southwest coast of Guatemala, Michael Coe and Kent Flannery (1967) identified a complex of salt-making sites which dated to the Late Formative period (Table 1.1). This discovery revealed saltmaking to be an industry of considerable antiquity in the Maya area; while surprising at the time, subsequent research has revealed evidence of saltmaking in several other Maya localities during Late Formative times.

Table 1.1.
General Chronology of Maya Area

1700	**COLONIAL PERIOD**
1600	
1500	
1400	**LATE POSTCLASSIC PERIOD**
1300	(Decadent Period)
1200	
1100	**EARLY POSTCLASSIC PERIOD** (Modified Florescent Period)
1000	
900	**TERMINAL CLASSIC PERIOD** (Pure Florescent Period)
800	
700	
600	**LATE CLASSIC PERIOD**
500	(Early Period II)
400	
300	
200	
100	**EARLY CLASSIC PERIOD**
A.D.	(Early Period I)
B.C.	
100	**LATE FORMATIVE PERIOD**
200	(Late Preclassic Period)
300	
400	
500	**MIDDLE FORMATIVE PERIOD**
600	(Middle Preclassic Period)
700	
800	- - - - - - - - - -
900	
1000	**EARLY FORMATIVE PERIOD**
1100	(Early Preclassic Period)
1200	

One region where early saltmaking has been documented is northern Yucatán. During a survey of the north and west coasts of the peninsula, Jack Eaton (1978) located many prehispanic salt-making sites, some of which also date to Late Formative times. Eaton's survey report, which includes a detailed study of the 2,000-year-old Yucatecan salt industry, will serve as a benchmark for future projects dealing with salt-making sites in Yucatán.

Saltmaking in the Maya area may have begun even before the Late Formative period. Evidence from the site of Pajón, on the south-central coast of Chiapas, raises the possibility that people there were engaged in saltmaking as early as the Early Dunas subphase, ca. 1000 B.C. (Paillés 1978). While the evidence is somewhat tenuous, it is nonetheless likely that such activities were taking place; after all, saltmaking in the Valley of Oaxaca, less than 500 kilometers away, began at an even earlier date (Drennan 1976).

Another major contribution to our knowledge of the Mayan salt industry is Lawrence Feldman's (1971) ethnohistoric survey of protohistoric trading patterns in southeastern Guatemala. Feldman has documented, in some detail, the location of the southeastern salt-making localities, the modes of production, the routes over which the salt was traded, and the trade range of each source. His dissertation is a comprehensive source of reference for this region of the Maya area, not only for information on salt, but for data on prehispanic trade in general.

In all these projects the investigation of prehispanic saltmaking was not the central focus of the research. One project of the late 1970s was, however, designed principally to investigate prehispanic saltmaking. This project was Brian Dillon's survey and excavations at the site of Salinas de los Nueve Cerros (known in Maya as Bolontehuitz) in the lowlands of the department of Alta Verapaz in Guatemala.

Salinas de los Nueve Cerros was a legendary salt source of the southern Maya lowlands. In fact, it was the only salt source of the southern lowlands and, consequently, a major source for that area. Dillon's investigations have revealed that the site next to the saltworks was occupied from Late Formative through the Classic period. The excavations have uncovered substantial remains of the ancient salt works and a variety of trade goods. Located on the Chixoy River, Nueve Cerros had easy access to the neighboring highlands and to the lowlands via the extensive Usumacinta River system. Given such a location, it is likely that Nueve Cerros was a prominent trading node between the highlands and the lowlands (Dillon 1977, 1979).

The final report of the Nueve Cerros project (publication date undetermined) should reveal substantial new information on the nature of highland-lowland trade. It will also shed new light on the role of the salt trade in the development of the Classic Maya civilization of the southern lowlands.

Few scholars have attempted to assess the economic and political impact of the salt trade on the development of Maya civilization. More than fifty years ago Mendizábal (1946a:330) suggested that the saltbeds of northern Yucatán were the primary objective of the central Mexican people who overwhelmed the peninsula in the tenth century. Without evidence to support his argument, however, it has been largely ignored. He perceived a salt source as the ultimate objective of every migration in Mexican history. Ironically, we must now recognize that he was, in part, correct. Recent archaeological evidence from the north coast of Yucatán suggests that many of the salt-making localities were occupied by the same foreigners who occupied Chichén Itzá in the Early Postclassic period (Ball and Eaton 1972; Ball 1977a; Andrews 1978b).

Salt has been perceived as a major factor in a number of events which shaped Maya history. E. Wyllys Andrews IV (1968; see also Ball 1977a, 1978; Andrews V 1981) believed that the Yucatecan saltbeds were a critical factor in the emergence of urbanism in the northern lowlands. He proposed that the rise of the city of Dzibilchaltún was mainly due to its control of salinas on the nearby north coast, and to the economic benefits it derived from the salt trade. It is probable that several north coast cities owe their existence to the salt trade, since this resource was the main northern export item in the voluminous trade between the northern and southern lowlands.

The cities of the southern lowlands were also affected by the salt trade, albeit differently. William L. Rathje (1971) has proposed that the need to import salt and other raw materials from distant sources was a seminal factor in the emergence of classic Maya civilization, and ultimately, one of the factors that led to its collapse. While Rathje's hypothesis has generated considerable debate among Mayanists, it has, on the whole, underscored the importance of the salt trade in the development of Maya civilization.

The questions and issues raised by researchers prompted my interest in the subject of salt. The first step was to inquire why salt was important to the Maya, and how and where they acquired it.

MAYAN USES OF SALT

To understand why salt was so critical to the Maya, we must explore the causes of imbalance in suppy and demand. The logical starting point is with the question "why the demand?"

Before the Spanish conquest, most of the salt produced in the Maya area was consumed in the diet. As a critical constituent of blood and other body fluids, salt is vital to human nutrition. Salt's essential ingredient, sodium (Na), is obtained by consuming sodium chloride (2.47 grams of NaCl provides 1 gram of sodium) or foods rich in sodium, such as meats. The human body demands that the concentration of sodium in the blood be kept constant, and a hormonal mechanism compensates for lack of salt by lowering its excretion rate in sweat and urine, while excess salt induces thirst. Uncontrolled salt intake can lead to a variety of medical problems, the best known being cardiovascular complications. Absence of salt leads to excessive secretion of water and, thus, to desiccation and death.

Although the human body needs a very small daily intake of salt to survive, the necessary amount varies considerably from person to person and region to region, depending on climate, diet, and working habits. The amounts recommended in medical and nutritional literature vary from 0.5 to 4 grams a day. These amounts are considered appropriate for the average sedentary North American citizen of the late twentieth century but are not necessarily applicable to people in other parts of the world.

People living in tropical climates have relatively high salt requirements. When involved in strenuous activities, their continual loss of salt through perspiration gives rise to even higher levels of consumption. This salt can be acquired through various indirect dietary means. However, when the vast majority of the population subsists primarily on a sodium-deficient diet of cereals and vegetables, the daily requirement can only be satisfied through consumption of mineral salt. According to medical and nutritional literature, the daily salt requirement of people living under conditions comparable to those of the Maya ranges from 8 to 10 grams, and in some cases, up to 30 grams a day. While a Maya farmer working in the fields may, in fact, require up to 30 grams a day, women, children, and infants would need considerably less. Thus, the average per capita consumption of the entire population probably falls in the lower end of the range, between 8 and 10 grams.

There is, in fact, substantial evidence in support of such a figure. The strongest corroboration lies in the statistics of the modern salt industry. Salt company officials in Mexico City, Mérida, Guatemala City, and San Salvador reported similar consumption figures: modern industrial statistics indicate a per capita consumption of approximately 10 grams a day throughout Mesoamerica.

Similar levels of consumption are reported in the ethnographic literature. At Chan Kom, Yucatán, a family of four consumed 12 kilograms of salt a year (Redfield and Villa Rojas 1934:57), a figure which translates to 8.2 grams per person daily. In the town of Baca, also in Yucatán, a per capita consumption of 8 grams a day has been reported (Arias García 1972:74).

Based on what storekeepers and housewives in a number of Maya communities told me, daily per capita consumption ranged between 8 and 12 grams. Most housewives, however, reported purchasing approximately 1 to 1.5 kilograms a month for families of four to six people: in short, their average per capita consumption was 8.3

grams daily. Using a different set of methods, Peterson (1976, 1977) came up with 8 grams a day as the mean per capita consumption of prehispanic Oaxaca.

It is possible to make some more general statements about salt consumption in the Maya area by using this figure. To begin, it is possible to calculate the salt needs of individual communities or regions for which we have population estimates, using the following formula: Population × 0.00292= X tons/year. Applying this formula to Tikal, which at its height in the Late Classic period is believed to have had a population of approximately 45,000 (Haviland 1970:193), the minimal need of that community would have been 131.4 tons a year. Since there are no salt sources anywhere near Tikal, this salt had to have been transported from distant regions. The logistics of transporting such large amounts are staggering and provide a glimpse of the complexity of the prehispanic salt trade.

Even more staggering are the estimates of salt needs at the regional levels. A region of major concern here is the Maya lowlands, which includes parts of Chiapas and Tabasco, the Yucatán peninsula, the Guatemalan Petén, Belize, the lower Motagua drainage, and adjacent lowlands of Honduras. Most scholars would now agree that at the height of the Classic period, around A.D. 800, the Maya lowlands had a population of five million or more. At 8 grams a day per capita, five million people would need 40 tons a day, or 14,600 tons a year.

Impressive as this figure may seem, it represents only a minimum estimate of the area's basic dietary needs and does not include the salt used for other purposes, of which the Maya had many.

Salt as a Preservative

Fishing was a major industry of the prehispanic Maya, and coastal dwellers traded fish to neighboring sites in the interior in exchange for agricultural goods. To preserve the fish, they either roasted, sun-dried, or salted it (Roys 1943:47; Lange 1971).

How widespread the practice of salting fish was in the Maya area before the conquest is uncertain. Salting was observed in the sixteenth century on the north coast of Yucatán (Landa in Tozzer 1941:190) and on the Pacific coast of Guatemala, though it was marginal there (Estrada and Niebla 1955:79). Even in this century in southwest Guatemala, McBryde (1947:79) observed that salted fish was available in highland markets only during Lent.

There is little archaeological evidence to support the notion that large quantities of fish were being shipped very far into the interior. Marine fish bones are rarely, if ever, found at sites in the highland or central lowland regions of the Maya area. Even at sites near the coast, fish remains appear in limited quantities: at Dzibilchaltún, 20 km inland from the north coast of Yucatán, only 5 percent of the vertebrate faunal sample was of marine origin (Wing and Steadman 1980). At Mayapán, 70 km inland, approximately 2 percent of the vertebrates were marine fishes (Pollock and Ray 1957:637). Although the excavations at these sites were not screened, the figures are low.

As a preservative, salt may also have been used in the tanning industry. Among the Maya, specialists were dedicated to tanning hides and pelts, which were used for clothing and other purposes. Many of these leather products were also traded (Tozzer 1941:89, 94, 121, 122; Roys 1943:46). We know almost nothing about how the Maya cured leather, or if they used salt in the process. In the Aztec capital of Tenochtitlán, Bernal Díaz del Castillo (1967:159 [Ch. 92]) observed large canoes full or urine, which was used for making salt and for tanning hides. Whether urine was employed by Maya tanners is also unknown.

Ritual and Medicinal Uses

Salt plays an important role in ritual and medicine throughout the Maya area. In fact, salt's ritual uses and the medical properties attributed to it are so numerous that only a brief discussion will be offered.

Several salt sources of the Maya area were, and in some cases still are, considered sacred places, pervaded by an aura of ritual. These include San Mateo Ixtatán and Sacapulas in Guatemala, Salinas Atzam in Chiapas, and Xtampú and possibly Emal in Yucatán, where the remains of prehispanic temple platforms, colonial or modern shrines, large wooden crosses, and other features clearly attest to the sacredness of these places. Remnants of ancient salt-related rituals still survive at some of these salinas, and though heavily syncretized with Catholic elements, they still retain traces of prehispanic religious practices. At San Mateo Ixtatán, for example, groups of elderly women burn copal incense in front of Catholic crosses while intoning chants that incorporate Maya and Catholic prayers. The purpose of this ritual is to invoke various deities—Catholic and Maya—to look over the salt springs and prevent them from drying up. The springs are, after all, a source of livelihood for many in the village (San Mateo supplies neighboring regions with salt), and the link between economics and religion is, in this case, quite real.

At Salinas Atzam, an old ornate Catholic church houses the Virgen del Rosario, the patron saint who looks after the tiny salt spring there. Atzam produces minute quantities of "sacred salt" used exclusively in ritual and medical practices in neighboring communities. The sanctity of Atzam and its salt is part of an intricate regional religious and sociopolitical system involving *santo* exchange and the maintenance of the Zinacanteco *cargo* system, as well as a host of other ritual activities.

The ritual use of salt in life-cycle ceremonies is widespread in the Maya area. Salt must be present at the beginning and end of life. Among the Zinacantecos of highland Chiapas, salt water is sprinkled around the inside of a house shortly after a birth or a death. At birth, a midwife offers some sacred salt to both parents. Throughout the Maya area it is common to place salt in the infant's mouth at the time of baptism, a European custom that may have been introduced by the Spanish.

Salt is also used in witchcraft rituals in both the highlands and lowlands—in exorcisms, in casting spells, and in impairing the power of witches (Redfield and Villa Rojas 1934:176, 179; Oakes 1951:163; Wasserstrom 1970:90).

Even the nonuse of salt was significant. Before certain festivals and on other appropriate occasions, the Maya would practice ritual salt abstinence. These fasts were common in both the lowland and highland regions (Landa in Tozzer 1941:107, 152, 158; Thompson 1966:69; Ximénez 1929-31, I:89 [Bk 1, Ch. 31]). It is not known why the Maya practiced such abstinence, or why, in certain contexts, they perceived salt to be a sacred item.

Possibly it is because of the value of this resource in many parts of the Maya area. Some regions had to import salt from great distances, which no doubt enhanced its economic and, thus, symbolic value. The practice of ritual abstinence also may be an unconscious commemoration of past salt famines, now long-forgotten events.

Salt's reputed medicinal properties may also explain its sacred status. At the village of Dzitás in northern Yucatán Robert and Margaret Redfield (1940:67, 72–73) reported several folk-healing practices involving salt—salt and oil were prescribed for epilepsy; marjoram, salt and the leaves of the *xul* tree (*Lonchocarpus* sp.) for preventing conception; and honey and salt for easing birth pains. At Zinacantán in the highlands of Chiapas, Vogt reported that "salt is rubbed on the patient's head in curing ceremonies; and salt water is used for divination and is held in the curer's mouth when he sucks on a patient" (1969:117). Many people in the western highlands of Guatemala told me that the salt from Sacapulas was an effective remedy for glaucoma. In fact, this belief is so widespread that friends in Guatemala City would often ask me to bring them salt from that source.

These are only a few instances of medicinal uses of salt. Many others exist, not only among the Maya but throughout Mesoamerica, and most are of prehispanic origin. Fray Bernardino de Sahagún, writing on the sixteenth-century Aztecs, for example, recorded a number of instances in which salt was used in healing (Sahagún 1975:589–94 [Bk. X]).

While information concerning the use of salt in folk healing is scanty, some evidence suggests that the Maya were conscious of the deleterious as well as salubrious effects of salt on a person's health.

McCullough noted that the Maya maintain ". . . a categorization of objects and processes, especially those related to foods and health, into 'hot' and 'cold' types, with an insistence on balancing the 'hot' and 'cold' in the body" (1973:32). Among the Yucatec Maya, salt is conceived as a "hot" substance (Redfield and Redfield 1940:64; Villa Rojas 1945:132; McCullough 1973:33), and it is taken to offset "cold" foods or illnesses. In a study of the "hot-cold" syndrome, McCullough has shown that salt intake, within the "rules" of the syndrome, represents an effective biocultural adaptation to thermal stress. As an example, he describes Maya workers laboring hard in their fields in the midday sun:

> The inevitable thirst requires water, but water, a "cold" substance, is thought to be dangerous. Taking a small amount of salt, a "hot" substance, with the water is felt to diminish the dangerous effect of the "cold" water on the man and to result in a more balanced *cuch* [thermal balance] (1973:33).

McCullough argues that this behavior is an excellent prophylactic measure against heat cramps, which are brought on by strenuous labor, excessive water intake, and salt depletion.

The hot-cold syndrome is also found in the highlands, but the perception of hot and cold appears to be different. Wasserstrom (1970:89) reports that among the Zinacantecos salt water is prescribed for fevers, since it is considered a "cold" medicine. The perception of salt as cold here is perhaps due to a behavioral adaptation to the colder climate and related stress factors of the highlands.

Another line of evidence that may help explain salt's attributed medical properties is its chemistry. The salt most prized for ritual use and medical properties came exclusively from highland sources. Analysis of salt from several sources revealed that, in addition to NaCl, the salt included other minerals, such as iron, calcium, magnesium, and various sulfates and nitrates. Possibly these minerals, rather than the NaCl, may be responsible for the healing properties attributed to the salt. A study of correlations between the minerals and specific medical properties and healing practices would be a valuable contribution to medical anthropology in the Maya area.

Salt as an Exchange Item

Given the importance of salt among the Maya, one would expect that it might have been used as currency, as in many parts of the world; but evidence suggesting such use in prehispanic Mesoamerica is scant.

I was able to locate only one report, a sixteenth-century document known as the *Relación de Meztitlán* from the central highlands of Mexico, in which salt is referred to as *moneda menuda* (small money) for making *cosas menudas* (small purchases) (Mendizábal 1946a:315). The term "small money" implies that salt was an item of small-scale barter rather than a unit of monetary currency. Historic and ethnographic reports from the Maya area corroborate this view.

In his history of Chiapas and Guatemala, completed in 1720, Fray Francisco Ximénez discusses barter, which was commonplace in Maya markets; among the items discussed was salt, which was exchanged for herbs and chile peppers (Ximénez 1929–1931, I:94 [Bk. 2, Ch. 33]). In a visit to the market at Rabinal in 1862, Habel observed this type of barter, noting that cacao beans, red peppers, leeks, soap and salt were exchangeable (1878:4).

Few traces of this system of barter survived into the twentieth century. McBryde (1933:124; 1947:84) found that salt had been used as "small change" in Sololá up until about 1900, and a team of ethnographers discovered that salt was exchanged for corn in Sacapulas as late as 1944 (Goubaud Carrera, González and Tax 1944:86). It was also used in small barter transactions in Ixtapa, Chiapas, in the 1950s (Zabala 1959).

Other Uses

Bishop Landa (in Tozzer 1941:35) noted that Maya warriors had quilted shields and jackets "made strong with salt and cotton," which offered effective protection against arrows. This body armor was introduced into Yucatán, along with the bow and arrow, from Central Mexico. To judge from available evidence, these innovations appeared in the Maya area in Late Postclassic times (Thompson 1966:147).

Finally, one case is known of the manufacture of salt artifacts. At the salinas of La Concordia in highland Chiapas saltworkers once made stars, crosses, and other items by allowing salt to congeal around reed frames. It is not known whether this practice dates to prehispanic times.

Post-Conquest Industrial Uses of Salt

During the early Colonial period, the Spaniards introduced new industries into Mesoamerica, several of which created a demand for vast amounts of salt.

The most voracious salt-consuming industry was silver mining. Shortly after the conquest, in the 1550s, a new amalgamation process for extracting silver from ore was invented in Mexico. Known as the patio process, the technique consisted of leaching silver-bearing ores with solutions of sodium. The resulting chemical reactions facilitated the extraction of silver even from ores with a low silver content. This process was employed throughout the Colonial period, surviving until the beginning of the present century. Since the silver industry was a major industrial enterprise, huge quantities of salt had to be shipped to the mines in the highlands of central Mexico. Local sources being inadequate, salt was transported from as far away as the Pacific coast and the northern coast of Yucatán. This demand had important growth consequences for the Yucatecan salt industry.

Silver was also mined in Honduras, and the introduction of the patio process in the seventeenth century transformed the primitive salt works along the shores of the Gulf of Fonseca into a major industry that still survives (MacLeod 1973:261).

Besides silver mining, two other Spanish industries gave rise to new uses for salt. Cattle ranching was a major occupation from the earliest days of the Colonial period in the Maya area, and the use of salt licks went hand-in-hand with this industry. As early as 1627 Friar Thomas Gage observed that mule owners brought their herds to Lake Amatitlán, where, for a fee, the animals were allowed to lick up the salt encrusted along the lake shores (1928:218–19). Other small salt works in marginal areas produced salt exclusively for cattle ranches. In his survey of Mexican salt sources, Zarate (1917) reported that many localities produced small quantities of low-grade salt, which was good only for cattle. Even today, the only thing that sustains many of the smaller salt works along the Pacific coast of Guatemala and Chiapas is the demand from nearby cattle ranches.

Little has been written about the tanning industry that developed along with the introduction of cattle, providing yet another use for salt. Of economic significance during the Colonial period, the industry utilized large quantities of salt (Suárez Molina 1977, I:323–27).

Modern Industrial Uses

The salt industries of Middle America have increased production enormously to satisfy a quantum leap in demand brought on primarily by the food and chemical industries. (Kaufman 1960 details modern industrial uses.)

Although the spread of refrigeration and modern transportation has rendered the practice of salting fish and meat almost obsolete, the uses of salt as a preservative have increased manyfold. Salt is now widely used in the preparation, packaging, and canning of foods throughout Mesoamerica, and the growing urban market for processed food can only increase.

Another major salt consumer is the chemical industries, in particular the petrochemical industries. Demand for the mineral has had a phenomenal impact on salt production in Mexico, especially in northern Yucatán, where salt production trebled in the 1970s; at present, approximately 50 percent of the salt produced is absorbed into industry.

The general industrial demand for salt has also stimulated salt production elsewhere. A large salt-making plant has been built in Coatzalcoalcos, Veracruz, and production from the salinas of Tehuantepec, Oaxaca, has increased dramatically.

In Central America, population growth and the spread of modern industry has fostered similar growth. In Guatemala, despite the wide-scale modernization and increasing productivity of the Pacific coast salt works, demand still exceeds supply, and sizeable quantities of salt are imported from El Salvador. This country has benefited enormously from its fast-growing salt industry, which not only supplies all internal national needs but also exports the mineral to neighboring countries. With the exception of Guatemala, these countries lack major salt industries (Belize has none), a fact which augers well for the Salvadorean industry.

SOURCES OF SALT

Salinas

The Maya acquired most of their salt from natural salt sources or salinas (fig.1.2). These consisted of coastal salt flats, inland brine springs, and salt playas exploited by a variety of techniques.

Most of the salt consumed by the Maya came from coastal salinas, primarily the Yucatecan salt flats located mostly on the north coast of the Yucatán peninsula, and the Pacific salinas, which extend along the coasts of Chiapas, Guatemala, and El Salvador from the Isthmus of Tehuantepec to the Gulf of Fonseca.

Salt-making methods differ from region to region, and in some areas the technology has undergone major changes in the last century. The most widespread method used today is solar evaporation, whereby salt water from coastal estuaries is collected in shallow pans and allowed to evaporate by solar action until only salt, known as *sal solar,* remains. Solar evaporation has been the standard salt-making technique of the Yucatecan salinas for more than 2,000 years; along the Pacific coast, on the other hand, the method has been applied only in the present century, with a few possible minor exceptions.

Saltmaking along the Pacific has traditionally involved filtering estuary water through large canoes full of salty marsh soil, which yields a saline brine. The brine is then cooked in large vessels, producing the final product, *sal cocida.* Until the last century, the brine was cooked in large ceramic vessels, which were replaced by large iron peroles or basins, which in turn gave way to shallow rectangular iron vats placed over crude adobe ovens. The sal cocida industry, which has also been in use for more than 2,000 years, is now being replaced by large modern solar operations. A few other methods, all variations on the two techniques, are used along the Pacific coast.

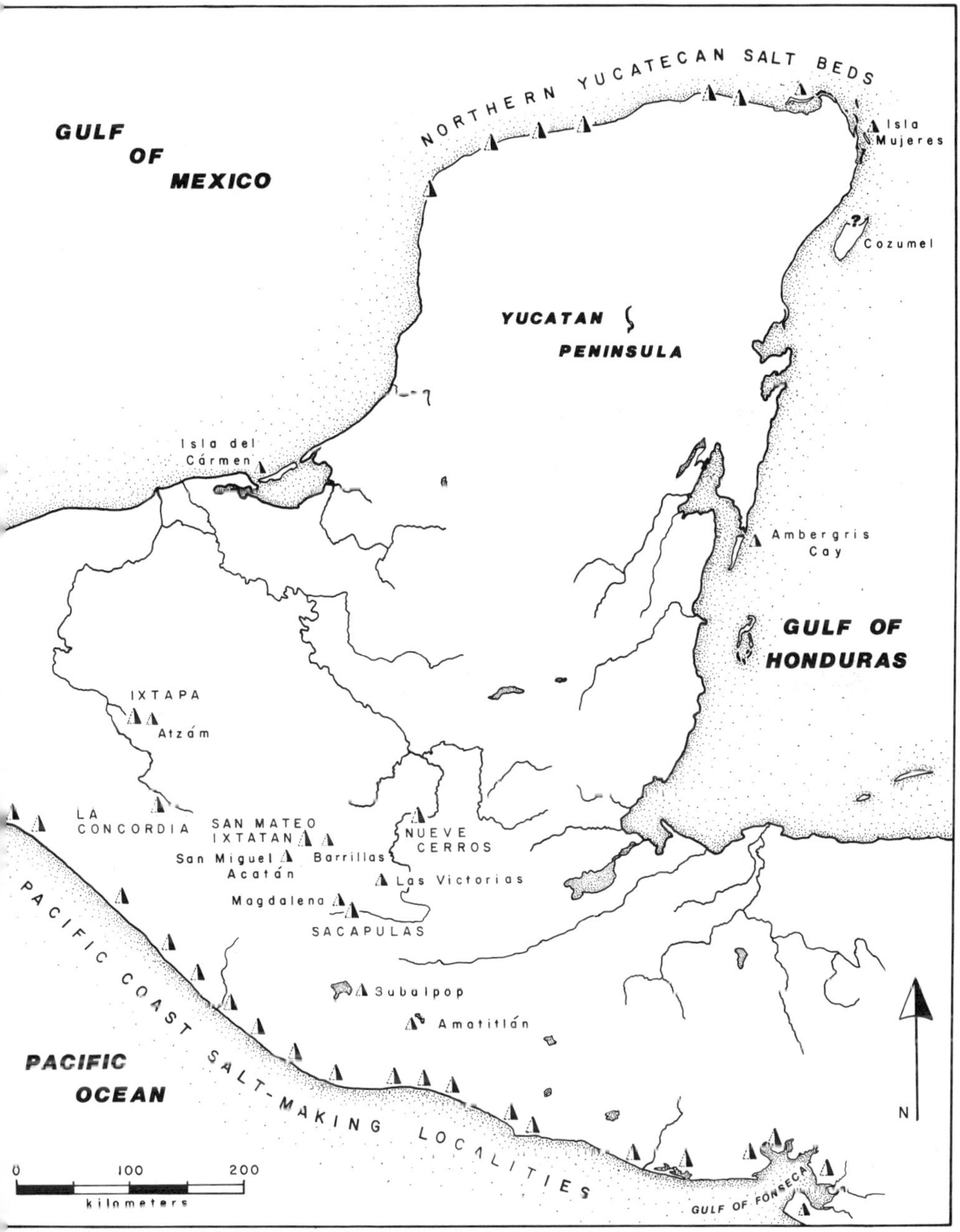

Fig. 1.2. Distribution of salt sources in the Maya area

With one exception, the inland salt sources are located in the highlands of Chiapas and Guatemala. Most of these are brine springs, whose salt is acquired by cooking. Methods of cooking vary, but the process is essentially the same in most places. At one locality, however, the brine was evaporated by solar action, and at two other places salt was acquired from salt playas by a method similar to the *sal cocida* technique used along the Pacific coast. The one source outside the highlands, Salinas de los Nueve Cerros in the Guatemalan department of Alta Verapaz, consists of a brine spring which flows out onto a salt playa. In the past salt brine was cooked here, possibly aided by solar evaporation.

Alternate Sources

In addition to the salinas, the Maya had several alternate sources of salt, including sea water, meat, and plants.

Sea Water. The salinity of sea water varies, but most marine waters contain an average of 3.5 percent salts in solution, of which 2.5 percent is sodium chloride (Multhauf 1978:7). Such a percentage is low when compared to coastal estuaries (up to 8 percent) or inland brine springs (often over 20 percent), which explains why sea water is rarely used as a source of salt.

Boiling sea water to procure salt is a laborious process which yields very small quantities of sodium chloride; it is generally considered a last resort, and only a few instances of it have been reported.

I was able to locate only two reports of sea water evaporation in the Maya area. One, an early seventeenth-century document in the *Archivo General de Indias* (AGI; Seville, Spain), notes that the now-extinct Toquequa Indians of southeastern Guatemala went to the nearby coast during the summer to make salt (Criado de Castillo 1605; Feldman 1975). Since there are no known salt flats or salt-producing estuaries near the Gulf of Honduras, they may have been boiling sea water.

The second report is an observation by Thomas Gann (1918:22) that the coastal Indians of northern Belize obtained salt in the early twentieth century by "evaporating" sea water. This report is dubious, at best, since local informants remember no such activity. Moreover, Gann may have been referring to the salinas on Ambergris Cay, in use at the time, which were exploited by solar evaporation.

Both these areas normally acquired their salt from northern Yucatán. However, at the times of the two reports, the northern trade had been cut off. If they were boiling sea water, it was a desperate response to a critical situation.

Animal Meat. As I have noted, salt can be indirectly procured through meat consumption, and Mary Pohl (1976:248) has suggested that the prehispanic Maya of the central and southern lowlands may have satisfied their basic requirements by this means.

This is a questionable argument, simply because one would have to consume substantial quantities of meat to satisfy the daily sodium requirement. For example, 100 grams of beef, pork, or turkey contain between 0.5 and 0.7 gram of sodium (Dauphinee 1960:403). To satisfy the Maya's basic sodium need, they would have had to consume between 370 and 500 grams of meat a day.

No evidence exists to indicate that the average Maya ever ate such quantities of meat. In fact, evidence suggests that the Maya traditionally had a low meat, protein-deficient diet (Roberts and Newman 1970:160-61; Scrimshaw and Tejada 1970:203). As little as 40 grams of meat a day have been reported consumed in northern Yucatán (Bonfil Batalla 1962:133–35), and ethnographic studies thoughout the Maya area attest to the scarcity of meat in recent times (Redfield and Villa Rojas 1934:38; Benedict and Steggerda 1936:184–85; Shattuck 1933:69; Arias García 1972:75; Pozas 1977:257–61; Valencia Bellavista 1978:116; McBryde 1947:10).

Some archaeological evidence confirms the ethnographic picture. Osteological studies of skeletons suggest a decrease in stature and an increase in nutritional stress among the Maya of Late and Terminal Classic and Postclassic times (Hooton 1940; Willey et al. 1965:535–39; Haviland 1967; Saul 1972, 1973), and zooarchaeological data convey the impression that meat consumption was very limited (Haviland 1965:17; Pohl 1976).

It would appear that the same situation prevailed in other parts of Mesoamerica. According to Sanders et al. (1979:488), meat was a scarce resource throughout the history of Teotihuacán. Analysis of the remains of white-tailed deer (*Odocoileus virginianus*) from sites in the basin of Mexico provides a rough idea of meat consumption through time in that area. The estimated maximum contribution to the diet is 30 percent during the Early Formative period, 6 percent during the Middle Formative, and 1.7 percent during the Late Formative; in all subsequent periods up to the conquest, it is less than 1.2 percent (Sanders et al. 1979:478). Thus, there is some evidence for significant levels of meat consumption prior to the Late Formative, but not for the subsequent periods.

It is likely that the Maya area underwent a similar pattern, with rising population density and deforestation (Deevey et al. 1979) leading to a decrease in game in Late Formative times. Thus, prior to this time, higher levels of meat consumption would have lessened the demand for mineral salt; then, with the Late Formative population buildup, the demand would have increased (Haviland 1975:26). Interestingly, large-scale salt production and trade appear to have begun at this time.

Plant Sources. Although most plants contain negligible amounts of sodium and chloride, a few species do contain appreciable amounts of salt or salt substitutes (such as potassium chloride), a fact that aboriginal populations in different regions of the world eventually discovered.

Where salt is scarce, certain species of palm, other trees, and grasses were exploited for this mineral. After the plants were burned, the ashes were soaked, and a salty residue was obtained through evaporation. This practice was common in certain areas of equatorial Africa, and North, Central, and South America. Several nineteenth-century travelers to West Africa's Sudan noted that the practice was so widespread that whole forests were denuded.

In the Maya area, two species of palms are known to have been exploited for salt. Fray Diego López de Cogolludo (1957:700 [Bk 12, Ch.7]), writing in the late seventeenth century, reported that in the Petén they burned a type of palm known as *xacsám,* which he also called a *guano* palm. We know now that he was referring unwittingly to two different species of palm, both of which were burned for salt: *Sabal mayarum* (also known as *S. mexicana, xan, botán, bayal,* or *guano),* and *Cyrosophila argentia* (also known as *Acantorrhiza moccini, kun, xocsám, or escobo).* Later reports also note the burning of guano and *botán* palms in the Petén and neighboring Belize (Gann 1918:22; Gann and Thompson 1931:326; Thompson 1965; Puleston 1978), and ethnographic studies have shown that the practice survived until recent times among the Lacandón Maya (Bruce 1968:119; Baer and Merrifield 1971:149, 220–21).

It is probably no accident that the reported geographical range of palm-burning, namely the central lowlands from Chiapas to Belize, coincides with the heartland of the Maya area, a region lacking mineral salt sources. It has been suggested that palms offered an alternate source of salt for the prehispanic populations of this region (Pohl 1976:247–48; Puleston 1978). This may have been the case prior to the Late Formative period, or after the Terminal Classic, when the area supported a small population, but it is unlikely that it was a significant alternate source during the Late Formative or Classic periods. Acquiring salt by burning palms is a laborious process which takes many hours and yields only a handful of salt for each palm tree (Jim Nations personal communication 1977). It is unlikely that such a source would have satisfied the requirements of the several million Maya of the central lowlands for a span of several centuries.

In parts of the world where salt was extracted from palms, the practice was generally confined to small, isolated tribes who lacked other sources of mineral salt. Clearly a last resort, the practice has been documented in several cases as a response to a decline in trade. The Maya area exemplifies such a case: the seventeenth-century Petén and Lacandón Maya were isolated tribal groups. The Lancandón burned palms until the turn of the century, when logging companies began supplying them with table salt; in northern Belize in the early twentieth century Gann (1918:22) explicitly notes that palm-burning was a response to a breakdown in trade with Yucatán, the area's main salt supplier.

Salt is such a critical resource that people have come up with some rather bizarre means of acquiring it. The Aztecs extracted it from urine (Díaz del Castillo 1967:159 [Ch. 92]), while the nineteenth-century Sumu Indians of east central Honduras would collect it by immersing burning sticks in the ocean and scraping off the tiny amounts of precipitated salt that adhered to them (Conzemius 1932:93; Stone 1966:221).

While the Maya did have alternate sources of salt, it is clear that these were of little consequence because their potential output is insignificant when viewed against the demand of several million people. The two main alternate sources of salt, animal meat and palm ash, may have satisfied the demand of early communities and nomadic groups prior to the Late Formative period. Thereafter, however, demographic pressure, deforestation, and a decline in game likely created a demand for mineral salt that outstripped supply.

The situation in the central lowlands during the Colonial period and nineteenth century may mirror the conditions of Formative times. Small isolated tribal groups appear to have survived quite well without an external source of salt, relying instead on game, dogs, turkeys and, when these did not suffice, on the ashes of palm trees.

The period that witnessed the growth of Classic Maya civilization saw a growing demand for salt which could only be satisfied through increasing production from the salinas, a pattern which in itself had serious implications for the development of Maya culture.

2. The Salinas of Yucatán

THE NORTHERN SALT BEDS

THE SALT BEDS OF NORTHERN YUCATÁN are the single largest source of salt in Mesoamerica. Located along the northwest and north coasts of the Yucatán peninsula (fig. 2.1), the beds lie underneath and along the edges of a chain of estuarine lagoons and swamps that were formed when the modern barrier beach emerged in early Holocene times.

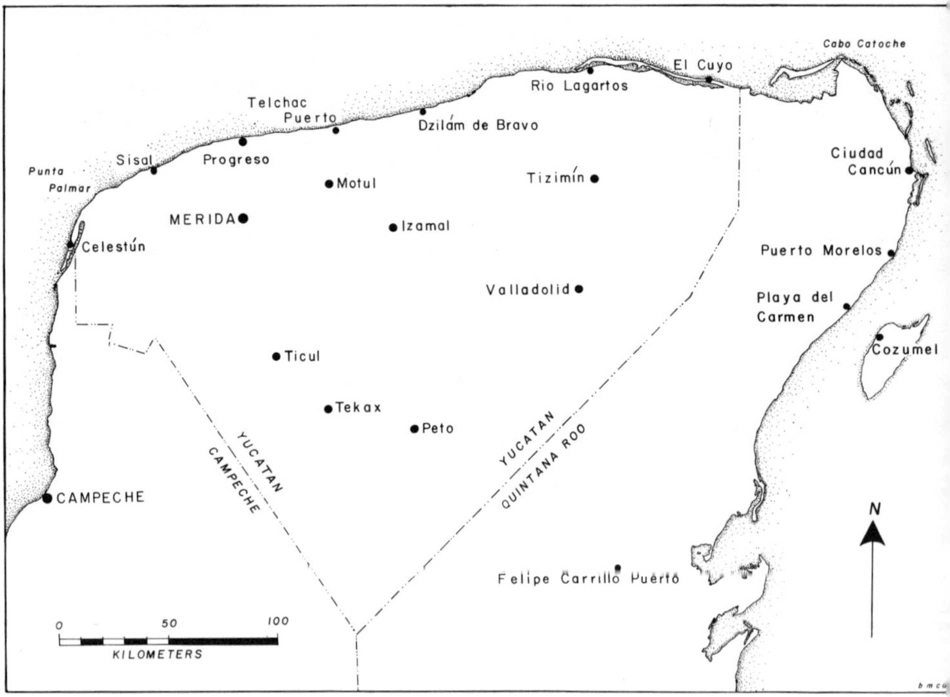

Fig. 2.1. Northern Yucatán

The water level of this coastal marshland zone, known as the *ciénaga*, depends on seasonal precipitation and runoff from the slightly higher adjoining interior. During the rainy season, from June or July through October or November, the ciénaga becomes a large inland sea, creating a natural barrier between the coastal strand and the interior. In the following dry months, from January to May, parts of the ciénaga dry up completely, and the lush swamp is transformed into a barrier landscape of parched, salt-encrusted mud.

Although the geological history of the Yucatecan littoral is complex and not well understood, the general sequence of events has been outlined by Eaton (1978:11), who conducted a survey of the north and west coasts of the peninsula in 1968:

> It is probable that the early barrier beach formations on the north coast of Yucatan, of which there are scant remains in the cienaga inland from the present beach, were formed during the past 3,000 years.
>
> The coastal survey has indicated that, in the main, the earlier habitation sites are found inland, situated upon what appear to be earlier beach ridges. The present barrier beach was the latest, and was probably formed from expanding sandbars by about A.D. 1000.
>
> There does not seem to have been steady beach progression along the north coast, where the most impressive barrier beaches are found. It appears that there were occasional intervals during which offshore sandbars developed until they formed effective barrier beaches, closing off lagoons behind them. The saline content of the water in the lagoons increased with repeated flooding and evaporation, precipitating rich deposits of sea salt.

Other accounts that discuss the geological history of the north coast and the formation of the saltbeds include the studies of Schott (1866), Urbina (1909), Edwards (1954), Collier (1964), Isphording (1975), and Wilson (1980).

The north coast saltbeds are a veritable gift of nature to man, as they offer almost infinite amounts of this valuable resource in exchange for a limited amount of labor. In most areas saltmaking is a seasonal activity, involving the simplest and perhaps oldest technique known to man: solar evaporation.

The salt works consist of grids of shallow rectilinear pans located along the shores of the lagoons (figs. 2.2, 2.3). At the beginning of the dry season, water which has been trapped in these pans starts to evaporate. In May, when most of it has evaporated and a thick layer of salt has formed in the pans, the harvest begins. The salt is simply raked up with wooden hoes and transported in baskets to the shores of the lagoons, where it is stored in large piles (fig. 2.4). In some cases, the harder salt is gathered by hand or, as was done formerly, with shell scrapers (Hester 1953:292). At the modern industrial salt works of Las Coloradas in northeast Yucatán, salt is no longer collected by hand but scraped up by large machines, dumped into trucks, and transported directly to the refinery. At the salt

Fig. 2.2. Aerial view of the salinas of Xtampú, Yucatán

Fig. 2.3. Salt pans at Chavihau, Yucatán

Fig. 2.4. Harvested salt at San Crisanto, Yucatán

works where manual laborers are still employed, the harvest may continue well into the rainy season, till June or July, because even though the salt pans are flooded, the layer of salt will remain solid for several weeks. In these operations shovels are used, and the salt is placed in small rowboats and carried ashore.

Saltmaking is an arduous task, and the average worker only puts in a few hours a day, most often from the first light at 4:30 or 5:00 a.m. until 9:00 or 10:00 in the morning. The harvest, which comes at a time of year when there is a lull in farm work (before the planting of corn), offers supplemental income to several thousand *milperos* (corn farmers) who make the annual pilgrimage to the coast.

Despite the seasonality of salt production, the mineral is easily stored, and thus is available year-round. Salt piles were once either covered with thatch or glazed with torches which hardened the surface against even the heaviest rains. I have not observed the thatch method since the late 1950s, and although the glazing method was still occasionally used in the mid-1970s, most of the salt is now stored in warehouses.

The Yucatán salinas are located at intervals along the coasts, and can be divided into three major groups: western, northwest and central, and eastern.

The western salinas are located on the Celestún peninsula, between Punta Desconocida and Punta Boxcohuo (fig. 2.5). These consist of several hundred salt pans spread over several thousand acres of land along a 24-km stretch of coast. It should be noted that although modern aerial photographs show over 700 pans, many are no longer in use. The pans vary in size from a few square meters to several thousand.

During the Colonial period, these salt works were known collectively as the Salinas de Campeche, since they were under the jurisdiction of that city. When the state of Campeche was created in 1858, the Celestún peninsula was bisected by the new state boundaries, leaving the southern half of the peninsula in the state of Campeche and the northern half in Yucatán.

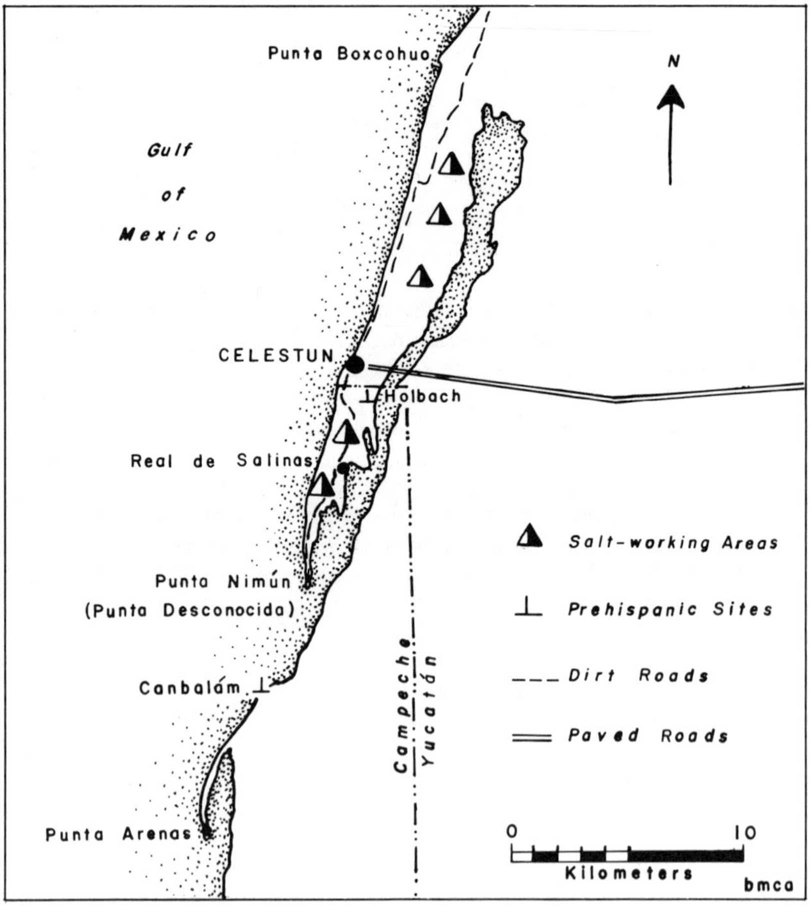

Fig. 2.5. The Celestún peninsula

The salinas of the northwest and central north coast are spread along 110 km of coast, from Punta Choventún in the west to Mina de Oro in the east (figs. 2.6, 2.7). Most of the salt works (including abandoned salt pans) are clustered around the following localities (names in parentheses are older historical names):

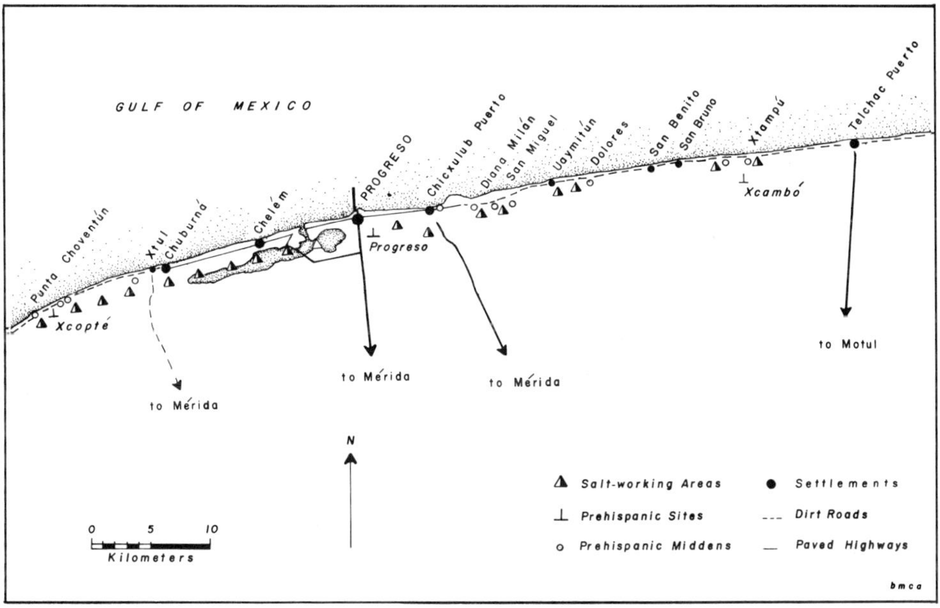

Fig. 2.6. Northwest coast of Yucatán

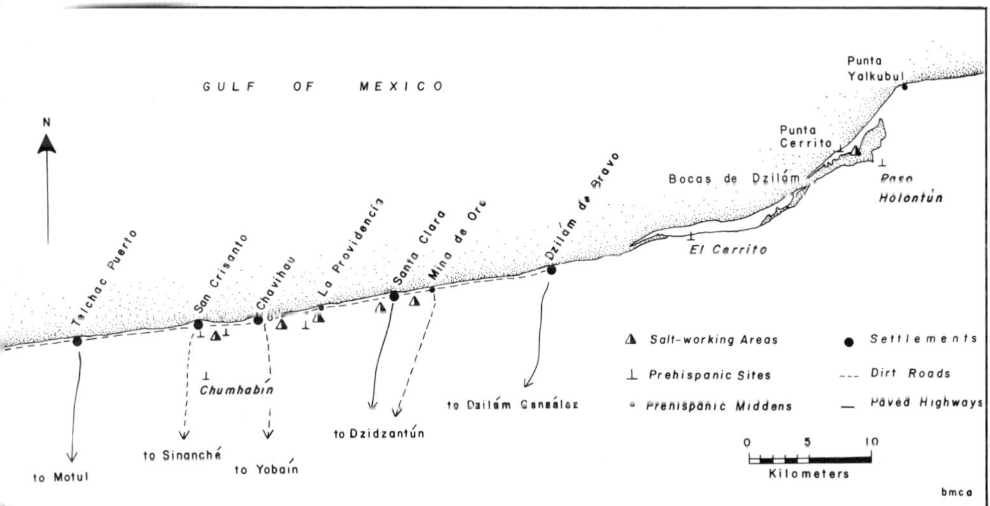

Fig. 2.7. Central north coast of Yucatán

1. Pta. Choventún (Rancho El Petén)—Xtul (Caucel beds)
2. Laguna Chuburná—Progreso [from the modern port of Chuburná to 2 km west of the port of Progreso; (Laguna de Chubulná beds)]
3. Progreso and Chicxulub (Chicxulub beds)
4. Diana Milán—San Miguel
5. Uaymitún—Dolores
6. San Bruno
7. Xtampú (Dzemul beds)
8. San Crisanto (Sinanche beds)
9. Chavihau—La Providencia (Yobaîn beds)
10. Santa Clara (Dzidzantún beds)
11. Mina de Oro (Dzidzantún beds)
12. Punta Cerrito

Most of these salinas are small operations; many are abandoned. Of the several hundred salt pans between Punta Choventún and Chicxulub, for example, only a few dozen were in operation in 1976. The salinas of Diana Milán—San Miguel, Uaymitún, and San Bruno are abandoned, and only a handful of the seventy salt pans at Xtampú were in use in 1976 (fig. 2.2). San Crisanto and Mina de Oro are the largest producers on the central coast today, and the salinas of Chavihau-La Providencia and Santa Clara are minor operations whose production has been declining since the 1960s.

The largest salt-making center in Yucatán is on the eastern north coast, halfway between the ports of Río Lagartos and El Cuyo (fig. 2.8). Known in the past as Emal and now as Las Coloradas, this salt-making complex consists of several square kilometers of salt

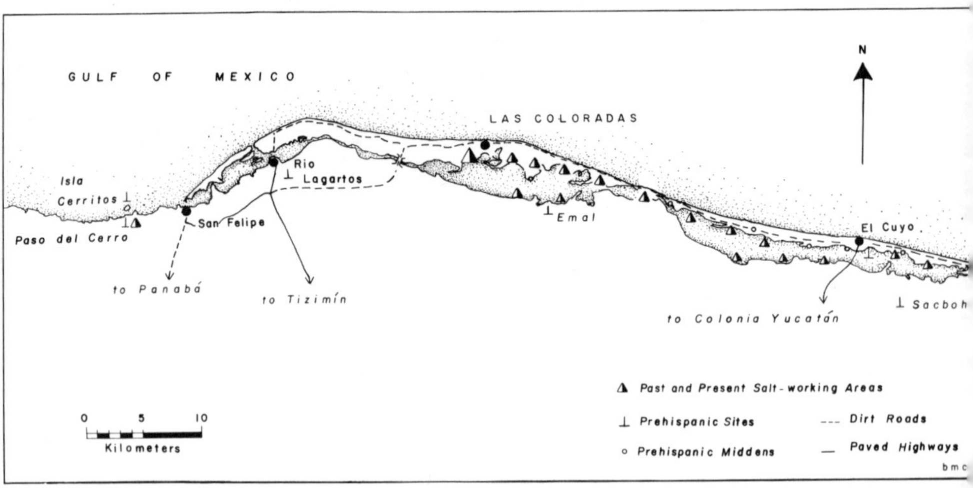

Fig. 2.8. Northeast coast of Yucatán

pans, a salt processing plant, warehouses, port facilities, and a salt company town (fig. 2.9). The operation is run by the Industria Salinera de Yucatán, S.A., a concern which has dominated the Yucatecan salt industry since the 1940s.

Fig. 2.9 Aerial view of Las Coloradas, Yucatán. In the 1970s this modern solar evaporation complex was the largest salt producer in the Maya area.

Unlike the other salinas of Yucatán, Las Coloradas operates year round. Water is pumped into large man-made lagoons from the adjoining estuary, the Río Lagartos. As evaporation begins, the water is channeled through a series of smaller lagoons until it becomes a highly saline brine. At this point it is pumped into shallow salt pans where the final solar evaporation takes place. It is then gathered by large salt-harvesting machines and trucked to the processing plant. The technology employed at Las Coloradas departs radically from the traditional salt-making methods used elsewhere on the peninsula and the results are dramatic: Las Coloradas now produces more than 80 percent of Yucatecan salt. Given the increasing demand for salt, it is likely that the major salt-producing localities on the northwest and north-central coasts will also be mechanized. While salt prices remain relatively stable, labor costs are increasing, and it is probable that smaller operations using manual labor will eventually go out of business.

Northeast Yucatán has long been a major salt-producing region, and the shores of the Río Lagartos estuary are littered with the remains of prehispanic and colonial salt-making sites. These sites are located along the ciénaga from Las Coloradas to several kilometers east of El Cuyo. In 1976, a few small traditional salt-making operations were still underway in this area.

In addition to these three major groups of salt works, a few isolated salt pans are scattered about the coastal swamps of northern Yucatán, many of them under water or covered by mangrove stands. It is unlikely that they will all ever be located, although in the course of my survey I found three that deserve mention.

At Punta Palmar (fig. 2.1) on the northwest corner of the peninsula, there is one large salt pan to the west of the lighthouse built by the lighthouse keeper during the 1950s. Three or four salt pans are located at Punta Cerrito, on the coast 25 km east of Dzilám de Bravo (fig. 2.7). Next to the salt pans is an archaeological site occupied during the Early Postclassic period, and the salt pans probably date to that period. Local informants said that the pans had been exploited in recent years but yielded only a few sacks of salt. Finally, a slightly larger complex of salt pans is located next to the site of Paso del Cerro 5 km west of San Felipe (fig. 2.8). Paso del Cerro appears to have been a salt-making village, occupied for more than 1,000 years, from Late Formative through Early Postclassic times. According to local informants, these pans have also been exploited in recent years.

Prehispanic History

Although saltmaking has been an important economic activity in northern Yucatán for at least 2,000 years, our understanding of prehispanic saltmaking is somewhat limited because of the nature of the data presently available. The archaeology of the north and the northwest coasts of Yucatán has been revealed in two surveys. Eaton's 1968 survey located most of the known archaeological sites along these coasts and identified a large number of prehispanic salt-making localities (Ball and Eaton 1972, Eaton 1978, Ball 1978). I conducted a second survey in 1976, the results of which have been partially reported (Andrews 1977, 1978a, 1978b; Garza T. de González and Kurjack 1980). Analysis of the surface collections has provided a preliminary chronological framework for the coast. Nonetheless, because none of the sites has been excavated, our knowledge of their time depth is extremely limited.

The earliest known evidence of human occupation on the north coast dates to the early phase of the Late Formative period, ca. 300 to 50 B.C. So far, this occupation appears to have been restricted to the Río Lagartos estuary of northeastern Yucatán, where Eaton (1978) located a cluster of middens, apparently the remains of small fishing and salt-making hamlets. Whether or not these are the earliest coastal sites in Yucatán is a problem of archaeological visibility: future investigations may reveal earlier occupations and evidence of earlier saltmaking, a not unlikely possibility given the presence of Middle and Early Formative occupations elsewhere on the peninsula (Hammond 1977, Velásquez Valadéz 1980, Andrews V et al. 1981).

During the following late phase of the Late Formative period (ca. 50 B.C. to A.D. 300), coastal sites appear all around the peninsula; in the saltbed zones between the Celestún peninsula and the El Cuyo region more than twenty sites with occupations date to this period. All of these sites lie in or adjacent to salt-making localities, and they would appear to represent hamlets or seasonal camps of fishermen and saltmakers.

While archaeological evidence suggests that saltmaking was a marginal industry in Late Formative times, some evidence indicates that north coast salt was being traded to other areas. Both Eaton (1978:48) and I encountered obsidian blades on the surface at three sites occupied exclusively during the Formative period. Since obsidian artifacts are rarely found at small coastal sites, the presence of these blades suggests trading activities. At this time, then, salt was probably being traded to nearby inland communities; however, there is indirect evidence that Yucatecan salt was also being traded to the distant southern lowlands during Late Formative times.

After the Formative period, many sites seem to have been abandoned. A few new settlements did appear, but the overall number of occupied sites declined. Only eleven sites with Early Classic occupations have been identified, and of these only seven are in salt-making localities. Whether this pattern represents a decline in population or a trend toward centralization is unclear (Ball 1978:129), for although sites dating to this period are scarce, they are much larger. In particular, a major settlement was constructed at San Crisanto on an island in the middle of the salt flats. Adjacent to this site are the remains of a large complex of prehispanic salt pans and circular platforms used for salt storage. There is also evidence of Early Classic occupation at nearby La Providencia, a site also located on an island in the midst of the ciénaga.

Farther east, on the Río Lagartos estuary, the sites of Emal and El Cuyo continued to be occupied, and although the evidence is not conclusive, it is likely that these sites emerged as the major salt-producing centers of the northeast coast.

Thus, while fewer sites appear to be inhabited during the Early Classic period, a certain degree of centralization of salt production at a few select sites located in the richest salt-producing areas is evident. Moreover, these sites, San Crisanto in particular, would appear to be producing salt in industrial quantities, intended no doubt for markets as far away as the southern lowlands. This shift from a Formative cottage industry to centralized large-scale production may have been a response to a growth in the volume of trade between north and south, a pattern also reflected in the appearance of southern-made polychrome ceramics in northern Yucatán. Some of these imported wares have even been found at the salt-making sites of La Providencia (Ball 1978:109) and El Cuyo.

During the Late and Terminal Classic period, the number of coastal sites increases slightly, but only a few appear to have been heavily occupied. Salt-making sites in the eastern zone, notably at Emal and El Cuyo, continue to be occupied, and the presence of imported Tepeu polychrome ceramics at El Cuyo hints at long-distance trade with the southern lowlands.

On the central north coast, La Providencia and Xcambó/Xtampú emerge as major salt-making centers. Both sites include small complexes of monumental architecture, and they were likely administrative centers or outposts of larger sites in the interior. An interesting feature of this period is the apparent absence of any Late or Terminal Classic remains at the large salt-making complex of San Crisanto; in fact, Eaton found no evidence of occupation following the Early Classic period. I doubt that this large complex of salt pans was abandoned, especially since they were reportedly in use in the sixteenth century. The lack of any evidence of occupation may be explained in several ways: it may be a problem of archaeological visibility, and future research may uncover remains from later periods; or the administrative and residential center of the salinas may have been shifted to La Providencia (5 km east) or Chumhabín (4 km south, in the interior). In the latter case, workers would have "commuted" to the salinas on a daily or weekly basis, leaving little or no trace of their activities. This "commuting" pattern appears to have been common during the Late Postclassic and Colonial periods, and I have observed it in recent years at several small salinas which are exploited by people from nearby inland villages.

The Early Postclassic period witnessed a dramatic change in the settlement patterns of the north coast of Yucatán. Several Classic period sites were abandoned while others appeared in new localities. The geographical location of Early Postclassic sites reveals an interesting pattern: they are located at strategic places, either on islands, at the mouths of rivers or estuaries, or next to important salt-making areas. As I have argued elsewhere (Andrews 1978b), the ceramics found at these sites suggest a deployment of foreign groups along the west and north coasts of the peninsula in the ninth and tenth centuries. By establishing themselves at key positions for trade and military purposes, these groups were able to take over the coastal trade networks and eventually gain control of the rich saltbeds of the north coast. By A.D. 1000 these groups, now known as the Itzá, had moved inland and established a capital at Chichén Itzá, from which they dominated northern Yucatán for two centuries.

Over the years, Chichén Itzá developed into a massive urban center, a centralized focus of power and wealth. Trade goods from distant regions of Mesoamerica and Central America flowed into the city, reflecting the importance of long-distance trade. Clearly, the Itzaes' control of the saltbeds and coastal trade networks provided the economic foundation for their political power, a situation they strove to consolidate.

Itzá outposts were established at many salinas: at Holbach on the Celestún peninsula, at Xcopté and Xcambó on the north central coast, and at Paso del Cerro, Emal, and El Cuyo on the northeast coast. There is also evidence of Itzá occupation at the mouth of every estuary along the north coast: at Canbalám, at the entrance of the Celestún estuary; at Punta Cerrito, next to the mouth of the Bocas de Dzilam estuary; and at Isla Cerritos, near the mouth of the Río Lagartos.

Isla Cerritos may have been the major port for Chichén Itzá. The island has excellent harbor facilities, the remains of a causeway connecting it to the mainland, and large quantities of Itzá-related ceramics (Tohil Plombate, Silho fine orange, Chichén Red and Chichén Slate).

Around A.D. 1200, the period of Itzá hegemony came to an end. An internal revolt led to the destruction of Chichén Itzá, abruptly ending the entire political and economic structure the Itzá had imposed on the region. Nowhere is this breakdown more evident than on the coast. Almost every site along the northwest and north shores of the peninsula was abandoned. Only Emal, the major salt-making site of the eastern salinas, shows evidence of significant Late Postclassic occupation. Traces of Late Postclassic remains have been found at a couple of other sites, but these do not suggest full-time occupation.

The abrupt abandonment of the northwest coast appears to have been a response to a shift in political and economic patterns. The Late Postclassic period witnessed a huge population buildup along the east coast of the peninsula, as well as a reoccupation of many sites in the interior which had been abandoned during Itzá times. The implications of these changes will be further discussed below.

Despite the absence of Late Postclassic communities on the northwest coast, there is good reason to believe that the saltbeds continued to be exploited. When the Spanish arrived in the sixteenth century, they encountered a large salt-making industry as well as a long-distance salt trade (cf. Roys 1943). Moreover, historical evidence shows that the rulers of the city of Mayapán, which succeeded Chichén Itzá as the political capital of northwest Yucatán from about A.D. 1250 to 1450, held control of the north coast salinas (Landa in Tozzer 1941:189; Gasper Antonio Chi in Tozzer 1941:230).

The near absence of Late Postclassic remains on the northwest coast is puzzling, particularly in view of the continued operation of the salinas. A partial explanation may be that of a "commuter" pattern. Sixteenth-century records refer to many salinas as belonging to neighboring towns in the interior: the Xtampú saltbeds are referred to as the Salinas de Dzemul, the San Crisanto beds as the Salinas de Sinanche, and so on (Roys 1957). Thus, during the Late Postclassic period, the salinas appear to have been exploited by itinerant seasonal workers who resided in the interior and came out to the coast on a daily or weekly basis. Future surveys will most likely encounter the campsites and middens that these workers left behind.

On the northeast coast definite evidence shows continued occupation at Emal, and it is likely that saltmaking there intensified in Late Postclassic times, as this would have been the major source for the large chain of communities that had developed along the east coast of Quintana Roo and farther south.

After the fall of Mayapán, around 1450, northern Yucatán was divided into a number of small provinces that were often at war with one another (Roys 1957). The coastal provinces controlled the saltbeds, and either supplied inland provinces with salt in exchange for agricultural goods or allowed groups from the interior access to the salinas in exchange for tribute. These coastal provinces also benefited from long-distance trade in salt. Particularly successful were the provinces of Ah Canul and Chauaca. Chauaca controlled the eastern salinas and exported most of the salt moving down the Caribbean coast of Quintana Roo and beyond to Belize and Honduras (Roys 1943, 1957; Thompson 1970).

Some confusion surrounds the tenure of the saltbeds. Although several sources report that they were communally held (López de Cogolludo 1957:180 [Bk. IV, Ch 3]; Molina Solís 1943 II:381), Bishop Landa has noted that Mayapán had a monopoly over the beds and that after the fall of Mayapán, a price was exacted by the province lords who were then in control (in Tozzer 1941:189). The notion of communal ownership of salinas probably was a legal fiction in the case of the larger, more profitable salt works that supplied the long-distance trade networks. Such a fiction would have made it easier for Spanish entrepreneurs to take over the largest salinas, and they did so shortly after the conquest.

Colonial and Modern History

When the Spaniards arrived in the sixteenth century, they were impressed by the size and extent of the Yucatecan salt works. Five principal sources of information on the industry date to that century, and all subsequent studies have relied almost exclusively on them.

The earliest account of the industry is that of Bishop Landa, written in 1566:

> There is a marsh in Yucatan worth recording, for it is more than seventy leagues long and is entirely salt. It begins from the coast of Ekab, which is near the Island of Mugeres, and con-

> tinues very near the seashore, between the coast itself and the woods, until near Campeche. . . . This marsh is saline for God has made there the best salt which I have ever seen in my life, since it is very white when ground, and as salt, those who know about it say that it is so good that half a celemin of it goes further than a whole one from anywhere else. Our Lord makes the salt in this marsh, there is a ridge of land, extending along its whole length, which divides it from the sea. In the season of rains then this marsh is filled up and the salt coagulates in the water itself in great and small lumps, which strongly resemble pieces of sugar candy. Four or five months after the rains are over, when the lagoon is somewhat dry, the Indians formerly had the custom of going to get salt, which they dry by collecting those lumps in the water and taking them away to dry. They had for this purpose their places marked out in the lagoon itself, which were the most productive of salt and with the least mud and water. And they were accustomed not to make this harvest of salt without the permission of the lords who in those places had most right by proximity. And to these all those who came for salt made some small offering, either of the salt itself or of things from their own lands; and since a principal named Francisco Euan, a native of the town of Caucel proved this, and also proved that the government of the city of Mayapan had given his ancestors charge of the coast and of the distribution of the salt, the Audiencia of Guatemala ordered those who should go to his territory to collect it at the present time to give him the same. A great deal is collected now at the proper season to be carried to Mexico and Honduras and Havana (in Tozzer 1941:189).

A briefer account was provided by Fray Antonio de Ciudad Real, who accompanied Fray Alonso Ponce in his tour of New Spain and Central America in 1588:

> On almost all that coast, from Campeche to the Rio de Lagartos and farther, there are wonderful salt-beds, which without being worked give quantities of coarse and very white salt of great value with which the whole province is provided, and ship-loads of it are carried to New Spain, Havana, Honduras and Panuco and other parts. These salt beds extend and stretch along the coast, following the seashore, and with rainwater when it falls, the salt coagulates in them and in the water they pile all the salt they can. Afterwards they take it from there and make large mounds on which they put fire which makes a thick and strong crust that does not melt although it rains for days and nights upon it; and if this care were not taken then it would melt and change into water. From those mounds [they make] loads and carry it inland, or they sell it to the ships that come for it (in Noyes 1932:307).

A more comprehensive source of information on the sixteenth-century salinas is the collection of documents known as the *Relaciones de Yucatán,* written by Spanish *encomenderos* (landlords) between 1579 and 1581 in response to a royal questionnaire. The questionnaire sought information on various topics, most important on the matter of available natural resources. Question 32 dealt specifically with salt, and twenty-nine relaciones commented on the abundance of salinas along the Yucatecan shores. A few contain details of ownership, production, and trade.

A fourth source of information is the 1549 Tax List (AGI Guatemala 128), a compilation of census and tribute data. Salt was among the tribute items, and all the towns that paid a salt tribute are listed therein.

The most valuable records of the early colonial industry are a collection of documents titled *Salinas de Yucatán,* which date from 1603 to 1608. Often referred to as the "1605 Salt Survey," these records contain surveys of the production of the Yucatecan salinas in 1603 and 1605 as well as correspondence related to legislation pertinent to the salt industry (AGI Mexico 72). These documents list all the salinas and the production of each.

After perusing the relaciones, the tax list, and the salt survey, it is possible to arrive at a reasonably good idea of the extent of the salt works in the sixteenth century. Almost every salt-making locality from the Celestún peninsula to El Cuyo appears to have been in operation, which was probably the case even before the Spanish arrived.

One feature that impressed the Spanish was the extent of the salt trade. At the time of conquest, Yucatecan salt was traded all along the Gulf of Mexico to the Pánuco River, and down the Caribbean shoreline to Honduras. Disappointed by the lack of precious mineral resources in Yucatán, the Spanish were quick to appreciate the value of the saltbeds; they took over some of the richest salt works, particularly on the northeast coast. Even though saltbeds were still, in theory, communally held, Spanish commercial interests prevailed and complaints of the natives went ignored (RY, II:215–16).

Royal authorities also took an interest in the salt works, and the king himself set about devising ways to derive revenues from the industry, not only in Yucatán but in other regions of the realm as well. As early as 1583, Phillip II was inquiring about the possibility of levying a tax on Yucatecan salt being shipped to Pánuco (Paso y Troncoso 1939–42, 12:125), and he was also considering ways to tax the Guatemalan salt industry (see Chapter 4). Finally, in 1591, Phillip decreed that all New World salinas be incorporated as property of the crown, and that a 10 percent tax be levied on all income from the salt works. The purpose of this was to raise funds for the building of an armada in the Pacific Ocean (AGCA 1591a, 1591b). This decree was not carried out until 1603, when his successor, Phillip III, ordered the Yucatecan saltbeds declared crown property. Surveys of the salinas were carried out at his behest in 1603 and 1605 (AGI Mexico 72). Local resistance to these orders, several bad harvests, and adminstrative problems led to a serious decline in productivity. Consequently, in 1609 the royal decree was rescinded, and the salinas reverted to communal status (AGI Mexico 72; López de Cogolludo 1957:456, 473 [Bk VIII, Ch 12; Bk IX, Ch 2]; Sarabia Viejo 1978).

Despite their failure to administer the salinas, royal authorities continued to tax the industry, often imposing harsh levies that undermined the competitive edge of Yucatecan salt in Mexican markets. In 1686, after pirates ransacked the city of Campeche, Charles II decreed that a stiff tax be laid on the salt industry to raise funds for massive fortifications to surround the city. Another tax was levied on Yucatecan salt imported into Veracruz. To make matters worse, in 1777 a vice-regal decree, probably designed to force down Yucatecan salt prices, gave preferential tariff rates to foreign salt, imported by way of Cuba into Veracruz. Finally, in 1790, a royal decree abolished the Veracruz tariffs, thus allowing the Yucatecans a larger share of the central Mexican market (Martínez Vera 1979).

We have few reports of the salt industry during the seventeenth and eighteenth centuries. In the course of his travels around the peninsula in 1675, William Dampier observed numerous salt works along the north coast and, in particular, around Celestún (1906, II:116–19, 144–46).

The only significant account during the eighteenth century was written in 1766 by two inspectors sent by royal authorities to assess the economic situation on the peninsula. Their report includes data on salt-making methods, production, and trade, as well as an exposé of widespread practices of fraud and tax evasion. Recommendations for improving the then-faltering industry were made, including a suggestion for more lenient taxes (Scholes et al. 1936–38, III:64–69).

Taxation and low levels of production continued to be the bane of the industry in the early nineteenth century. Moreover, exports to Veracruz and central Mexico, which were substantial prior to 1810, ceased with the War of Independence. Another matter which frustrated the Yucatecans was their inability to corner the Cuban market: Yucatán occasionally shipped salt to Havana, but the Cubans apparently relied mostly on U.S., British, Spanish, and Portuguese imports. Other events that undoubtedly affected production and trade were the separation of Yucatán from Mexico (1840–42) and the Caste War of Yucatán of 1847–51).

Despite these adverse circumstances, the industry survived. In 1845 salt was being shipped to nine Mexican ports on the Gulf of Mexico and to Belize (Regil and Peón 1853). After the Caste War, the industry resumed and increased over the years. Salt was shipped to Mexican ports as well as to Belize, Cuba, and even Liverpool, England. After the conclusion of the War of Independence, and throughout the rest of the century, the bulk of the exported salt went to central Mexico, where it was used in the silver refining process.

The eastern salinas were undoubtedly abandoned during the Caste War, but were later revived. By the late 1870s production was again under way, and within a century they were again the hub of the Yucatecan salt industry.

There are several accounts of the nineteenth century salt industry. The most extensive are those of Regil (1848–49), Echánove and Bolio (1849), Regil and Peón (1853), and Suárez Molina (1977).

Toward the end of nineteenth and beginning of the twentieth centuries, Yucatecan production fell. Four reasons account for this occurrence: (1) a new silver-refining process which no longer required salt had been developed; (2) the Mexican revolution of 1910 disrupted the mining industry as well as trade for more than a decade, curtailing salt imports to Veracruz; (3) toward the end of the century, shipments to Cuba appear to have ceased, perhaps because of the Spanish-American War of 1898; and (4) shortly after the turn of the century, even the trade with Belize broke down. Gann (1918:22) reported that "owing to the troubles with Mexico" the salt supplies had been almost entirely cut off.

By this time, the breakdown in the trade to the southeast (Quintana Roo and Belize) was apparently complete. Belize was receiving shipments of salt from Britain, some of which found their way into southern Quintana Roo, the territory of the rebel Icaiché and Santa Cruz Indians (Grant Jones personal communication 1979).

The recovery of the industry began in the late 1930s, when two brothers, Joaquín and Roger Roche, began to revitalize various north coast salt works. In 1946 they created the Industria Salinera de Yucatán, S.A., and built the first salt refinery on the peninsula at Las Coloradas on the northeast coast. This refinery produced the well-known "Sal Sol" brand of table salt, which was marketed throughout southern Mexico.

Several factors have spurred the revival of the industry during the twentieth century. Massive population growth has created an increasing market for table salt, while the cattle, packaging, and petrochemical industries have also created new demand. The Roche family and several other minor producers brought several salt works back into production, mainly those of Celestún, Chuburná, Chicxulub, Xtampú, San Crisanto, Chavihau, Santa Clara, Mina de Oro, Las Coloradas, and El Cuyo. As in earlier times, Celestún and Las Coloradas have become the main producers. The Celestún salt works, operated during the nineteenth century by the Ferrer family of Campeche, are now exploited by a number of producers.

The Roche family used to exploit a number of salinas in different localities but have concentrated their efforts at Las Coloradas, where they have built the largest salt-making complex in Mesoamerica. Their production has increased tenfold: in 1980 they produced close to 500,000 tons, the majority of the salt on the peninsula.

There are several twentieth-century accounts of the Yucatecan salt industry, including the reports of Zaráte (1917:30–31), Lozano García (1946:43, 59–60), Hester (1953:292), Edwards (1954), Triay Peniche (1977), and Eaton (1978).

Production

Throughout history, Yucatán has been the largest producer of salt in Mesoamerica: in fact, its production remains several times greater than that of other Mesoamerican salt-making regions. The following figures for 1972 provide an idea of the size of the industry in relation to other regions: El Salvador, the second largest salt-making region in the Maya area, produced 37,000 tons (DGEYC, AE 1972), while the salinas of the Pacific coast of Guatemala (the third largest producer) yielded 11,000 tons (RG, DGE n.d.); Yucatán, meanwhile, produced approximately 200,000 tons (J. Roche personal communication 1976).

Although we have no production figures for prehispanic times, the data from the 1603 and 1605 surveys provide an approximate estimate of production shortly after the conquest (AGI Mexico 72; see Appendix A for a tabulation and interpretation of these data). Depending on the way in which figures are interpreted, the reported production of 1603 to 1605 ranged between 17,000 and 19,000 tons. For reasons discussed in the appendix, I believe these figures to be very conservative: 20,000 tons is probably a more reliable minimum estimate. Nonetheless, the amount of salt reported was sufficient to have supplied the basic needs of six million or more people—enough to have met the demand of the population of the Maya lowlands throughout history.

Even though the available records do not fully document such a case, Yucatán probably maintained this level of production throughout the Colonial period and most of the nineteenth century. The next set of figures date from 1761 to 1765 and refer only to tax records of salt exported from the port of Campeche to Veracruz. The figures range from 7,000 *fanegas* (805 tons) in 1761 to 13,950 fanegas (1,604 tons) in 1765 (Scholes et al. 1936–38, III:67; a fanega weighs 115 kg: for an explanation, see Appendix A). However, the same report notes that tax evasion was widespread, so the figures represent only a part of the exports and a small fraction of overall production.

Slightly higher export figures are reported for the early nineteenth century: An average of 18,500 fanegas (2,127 tons) a year between 1805 and 1809 (Quirós 1973:251), and 25,000 to 30,000 fanegas (2,875 to 3,450 tons) in 1811 (Regil 1848–49, No.7). All these figures are based on tax records and pertain only to salt exported to Veracruz; they do not include shipments to other ports in the Gulf of Mexico, Cuba, or Belize.

A report of 1814 records the production of the Yucatecan industry at 80,000 to 100,000 fanegas (9,200 to 11,500 tons) a year (Echánove y Bolio 1849, No. 28). Though undoubtedly a conservative figure based on tax records—the only source of such information—it does reflect a decline in production brought on by the Mexican War of Independence. The report notes that shipments to central Mexico had ceased, depriving the industry of its major client.

In 1845, just before the Caste War, Yucatán reportedly produced 80,000 fanegas (9,200 tons) (Regil and Peón 1853). This figure, also based on tax records, is undoubtedly very low and unreliable.

The early nineteenth century production figures are unreliable because they are based on tax records from an era in which smuggling, malfeasance, and tax evasion were widespread (Suárez Molina 1977, II:33–34, 112–13, 267–69).

For the late nineteenth and early twentieth centuries data are more reliable (Zaráte 1917:29–30). At the Real de Salinas alone, he reports that production ranged from 16,000 to 23,000 tons a year; he does not report a specific figure for the rest of the salinas, but cites an 1881 report which records that the surplus carried over from one year to the next ranged from 9,000 to 12,000 tons. Moreover, the report notes that the estimated potential production of the entire industry was 57,500 tons. This is an interesting figure, as it suggests the potential of the industry long before the advent of modern technology. Since the technology of the industry in 1881 was comparable to that at the time of conquest, the only difference being transport facilities, the prehispanic Maya could have produced a similar amount. Figure 2.10 summarizes Yucatecan salt production for four centuries.

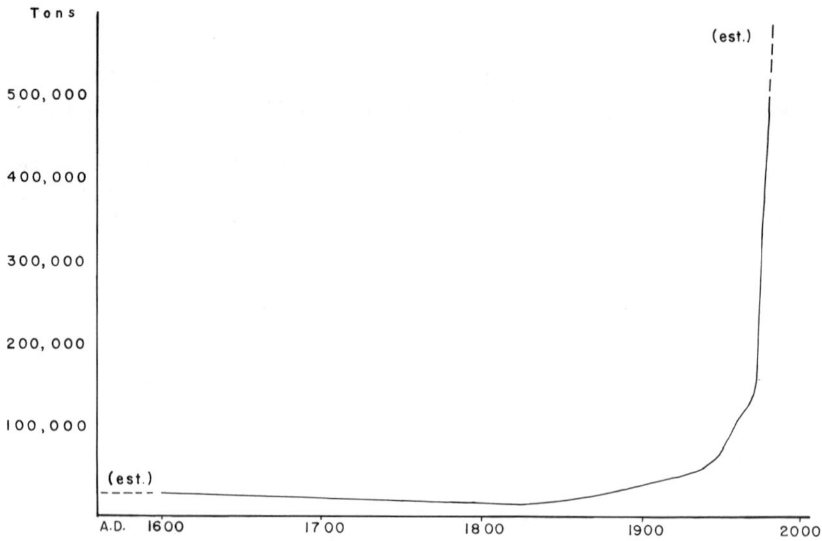

Fig. 2.10. Salt production of Yucatán from 1600 to 1980

In view of the growth of Mexican industry, the increasing demand for industrial salt will undoubtedly continue, with ample room for expansion. According to one estimate, the potential production of the north coast saltbeds is more than 2.5 million tons a year (Solís Preciat 1969:45).

Trade

When the Spanish arrived, they encountered a salt trade that spread from the north coast of Yucatán to three distinct geographical regions. The first was the local market, namely the communities of the northern part of the peninsula. That communities throughout Yucatán were supplied with salt by overland trade is well documented in early colonial accounts (RY; Roys 1943, 1957; Piña Chan 1978). This internal trade began in Formative times and continues to this day.

The second axis of the salt flow is the "western trade." Its salt came mainly from the Celestún peninsula, and probably from some of the salinas between Sisal and Progreso on the northwest coast. It moved by boat down the west coast of the peninsula and beyond to the Gulf Coast lowlands. By the sixteenth century, the salt had reached as far as the Pánuco River in northern Veracruz, a distance of more than 600 miles (Thompson 1966:221).

The western trade also supplied the base of the peninsula and the western lowlands via several river drainages, notably the Candelaria, Usumacinta, San Pedro Mártir, and Grijalva rivers. This trade reached deep into the Petén: at the time of conquest, Yucatecan salt was being traded as far inland as the Cehache territory (RY,I:369; Roys 1943:52–53; Scholes and Roys 1948:59, 244).

The third axis was the "eastern trade" in which salt from the northeastern salinas was traded down the Caribbean coast as far as the Gulf of Honduras. This trade also supplied inland areas at the base of the peninsula via several waterways, namely the Hondo, New, Belize, Sarstoon, and Dulce rivers. Through these networks, Yucatecan salt reached into the Petén and possibly into the eastern highlands of Guatemala via the lower Motagua basin (Roys 1943:56; Thompson 1970:129, 149–50; Viana et al. 1955:21).

The basic structure of the Yucatecan salt trading network probably existed from Formative times until the early years of the twentieth century. However, throughout the 2,000 years of its existence, the trading network underwent a series of drastic shifts and interruptions, responses to political and economic change.

YUCATAN: MINOR SOURCES

In addition to the north coast saltbeds, there are four and possibly five other salt sources on the Yucatán peninsula. These sources, no longer in use today, are also natural saltbeds exploited by means of solar evaporation. All of these salinas are very small, and were probably drawn on for local consumption only. Together, they probably had a potential production of less than a thousand tons a year. They were located at Isla del Cármen, Campeche; Isla Holbox and Isla Mujeres, Quintana Roo; and Ambergris Cay, Belize. A possible fifth salina may have existed on Cozumel Island, Quintana Roo.

Isla del Cármen, Campeche

Several sources allude to the existence, in the past, of a small salina on Isla del Cármen (Zaráte 1917:30; Lozano García 1946:43; Lefond 1969:132-34). It appears to have been in use in the nineteenth century, for tax records show that 172 fanegas (20 tons) were shipped from Cármen to Tabasco and Coatzacoalcos in 1845 (Regil and Péon 1853).

There is also some evidence that these salinas were in use in prehispanic times. An early account refers to some prehispanic salinas on the western side of the Laguna de Términos (RY, I:354; see also Roys 1943:105; Ruz Lhuillier 1969:39). Since I was unable to locate any salinas in the area other than those of Isla del Cármen, it is likely that the report refers to the saltbeds on the island. The report suggests that these salt works, which once yielded a fair amount of salt, were abandoned by the last half of the sixteenth century.

Local informants had only vague recollections of the salinas at Cármen, which were located on the western tip of the island. In recent years the saltbeds have been covered by the modern city of Ciudad del Cármen.

Zaráte (1917:30) also mentions salinas at nearby Champotón, a report later cited by Mendizábal (1946a:298). However, there are no saltbeds at Champotón, and Zaráte's report undoubtedly refers to the salinas of Isla del Cármen, which were under the jurisdiction of Champotón.

Isla Holbox, Quintana Roo

Just south of the town of Holbox, near the northeast corner of the peninsula (fig. 2.11) are two small salt pans. According to local informants, these pans have been in use in modern times and produced approximately 10 tons a year. They were important to the inhabitants of Holbox, most of whom are fishermen, because their fish had to be preserved by salting.

The pans are very small, and we know that the yield did not satisfy even local demand, for the fishermen still had to import salt from nearby El Cuyo.

In the early 1960s the fishermen began to refrigerate their catch, and the beds were abandoned. Today, the beds are partially covered by mangrove plants which soon will erase all trace of them.

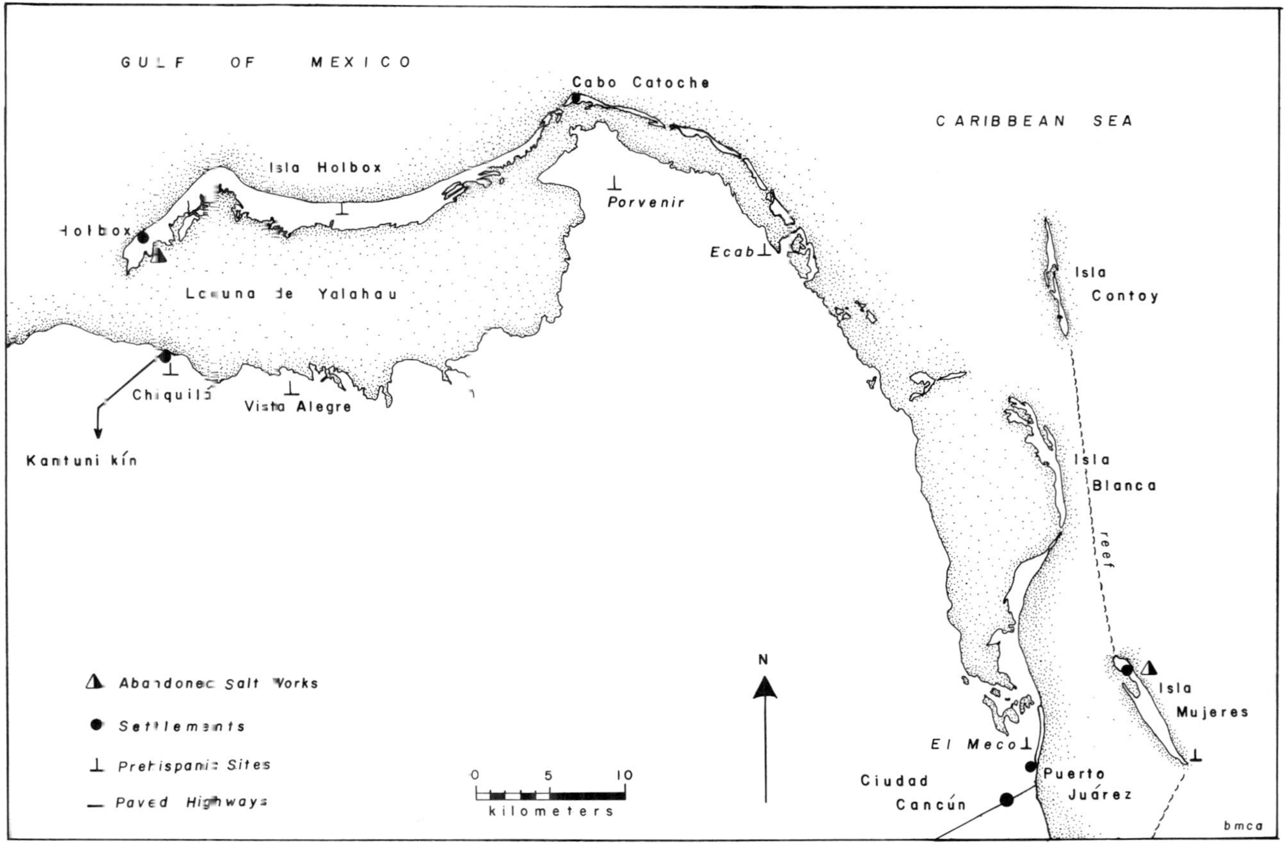

Fig. 2.11. Northeast coast of the Yucatán peninsula, showing the Isla Holbox and Isla Mujeres salinas

Saltmaking appears to have been a recent activity of the modern inhabitants who settled on Holbox in the late nineteenth century.

Isla Mujeres, Quintana Roo

Isla Mujeres was the first place where the Spaniards saw native salt works on the American mainland (fig. 2.11). Bernal Díaz del Castillo, who accompanied Cortés on his visit to Isla Mujeres in 1519, reports that the natives grew corn and made salt (Díaz del Castillo 1967:45).

Shortly after the conquest, the island was abandoned. A report of 1579 (RY, II:173) notes that the salinas had been abandoned; however, the 1605 Salt Survey notes that natives from the mainland port of Ecab went to the island to harvest salt (AGI Mexico 72).

Subsequent to this report, there is a 200-year gap in the history of the island and no reference to it appears until the 1840s, when it was reoccupied. Shortly after the reoccupation, the salinas were again exploited (Regil and Peón 1853:269). In a visit to the island with her husband in 1877, Alice Le Plongeon (1889:14–15) encountered a small, active salt industry there; her account of it is the only firsthand description of the salt works prior to the twentieth century.

Following Le Plongeon's account, we have only passing references to the salt works. Zaráte (1917:31) mentions them, as do the studies of Mendizábal (1946a:299–300), Roys (1943:47, 1957:145), Lozano García (1946:map), Edwards (1954:16), Thompson (1966:218, 1970:129) and Lefond (1969:132–34).

The salinas of Isla Mujeres consist of three sets of salt pans, just south of the modern town, known as Salinas, Salinitas, and Salinas del Cañotal. The last is the largest, measuring approximately 100 by 500 meters. The mode of production was the same as that in northern Yucatán. One unusual feature reported by Le Plongeon—and of potential interest to archaeologists excavating salt-making sites—was the use of turtle shells rather than baskets to carry salt. As in Holbox, most of the salt was used for salting fish and for satisfying local needs. Isla Mujeres never exported salt in any significant quantities, though local informants recall an occasional shipment to Cozumel.

In fact, it would seem that Isla Mujeres did not have the potential to export any significant quantities of salt. The 1605 Salt Survey records a production of 7,000 fanegas (805 tons; AGI Mexico 72), a figure suspect for several reasons. The first is that the figure cited obviously is only an estimate made by a bureaucrat who did not ever go to Isla Mujeres. The language used is a dead give-away: " . . . they probably collect . . . ." Second, my own inspection of the salinas leads me to question such a figure: they are simply too small to yield such a quantity of salt. Joaquín Roche (personal communication 1974), one of the leaders of the Yucatecan salt industry, agrees. He estimates the maximum potential production of the salinas to be

100 tons a year. And finally, my main informant on Isla Mujeres, Don Ausencio Magaña, laughed at such a figure. He is an older man, familiar with the salt production of the early twentieth century. According to him, annual production was about 100 tons a year; maximum potential production, under the best of circumstances, may have been approximately 150 tons a year (Magaña personal communication 1977).

Production began to decline in the late 1940s and early 1950s, when refrigeration units began arriving on the island. In the late 1950s, the opening of the Valladolid–Puerto Juárez highway and daily ferry service to the island put an end to the industry. Cheap commercial salt from the mainland had become available, underselling the local product.

Although a few families continued to harvest salt for their own needs, scraping up a "natural" harvest at the peak of the dry season, the pans were not kept up. Eventually, Salinas and Salinitas were abandoned, and by 1970 had become garbage dumps. Salinas del Cañotal, the last salina to be exploited, saw its last harvest in 1974 (Magaña personal communication 1977). Today it is an abandoned marsh.

Isla Cozumel, Quintana Roo

Several authors have reported that Cozumel Island once had salt works (Lozano García 1946:map; Lefond 1969:132–34; Freidel 1976:381; Piña Chan 1978:40). Whether Cozumel ever produced salt is debatable. There is a series of shallow lagoons on the north coast of the island; and one, next to the prehispanic site of Aguada Grande, has a few submerged walls that may have enclosed salt pans in the Late Postclassic period (Freidel 1976:230).

Local informants have no recollection of saltmaking on the island (I have been inquiring since the early 1970s), and find such a notion unlikely. In the 1940s, Ausencio Magaña of Isla Mujeres visited the northern lagoons of Cozumel to assess their saltmaking potential and concluded that they could not yield any significant quantitites of salt (Magaña personal communication 1977).

While there appear to be salt deposits beneath the lagoons, the water rarely evaporates, and since the lagoons are open to the sea, tidal action would prohibit saltmaking. If a sandbar had closed off the lagoon during prehispanic times, however, saltmaking would have been possible. The potential salt-making area is small, and I would concur with Freidel (personal communication 1977) that the likelihood of salt having been produced in any significant quantity is very slim.

One of the documents that led scholars to believe there were salinas on Cozumel is the 1549 Tribute list, which lists Cozumel as having paid a tribute of six fanegas of salt in that year (AGI

Guatemala 128, Roys 1957:155). Such a payment is not surprising, as other communities without salinas or access to them also paid salt tribute. Such was the case of Polé (Xcaret) and Zama (Tulum) on the mainland opposite Cozumel. As Thompson notes (1970:129), the salt was probably gathered elsewhere.

Thus, I have found no solid evidence of salt production in the historic period. Lozano García's (1946:map) report is unsubstantiated, and Lefond's (1969:132–34) is based on Lozano García's. Piña Chan (1978:40) bases his report of salinas on the 1549 salt tribute report, which does not necessarily imply production. In fact, the several early colonial accounts of Cozumel fail to mention saltmaking.

Ambergris Cay, Belize

The only known salt source in Belize is located at the northern end of Ambergris Cay, just south of the Mexican border at Boca Bacalar Chico (fig. 1.2). Local informants report that two salt lagoons that dry up during the dry season were once harvested.

These salinas were exploited by local people from the town of San Pedro and by Mexicans from the nearby port of Xkalak. Occasionally, fishermen from the town of Sarteneja, on the nearby Belizean mainland, also came to the salinas to harvest salt. Sometime around 1960 the lagoons stopped producing salt. Some local informants say the beds were "exhausted," but a rise in annual precipitation seems more likely. The lagoons are generally flooded nowadays, even during the dry season.

Production estimates are hard to find since the salinas were exploited by individuals who collected only a few sacks on each of their sporadic visits. One informant recalled a major excursion in which several individuals from San Pedro collected almost a ton of salt.

Most local estimates of the potential production of the two lagoons ranged between 10 and 50 tons a year; one fisherman suggested 100 tons. I did not actually see the salinas, since they were flooded at the time of my visit; however, a brief aerial reconnaissance leads me to believe that the two lagoons could not yield more than 50 tons a year.

The salt from these salinas appear to have supplied local populations, mainly at San Pedro and Xkalak. Small quantities were traded to Sarteneja, and to Cay Caulker, the next island to the south, but that appears to have been the extent of its distribution

I assume these salinas were in use in prehispanic times, as local informants report a number of prehispanic middens adjacent to the two lagoons. Moreover, nearby prehispanic sites on the cay are known to have remains dating to the Early and Late Postclassic periods (Peter Schmidt personal communication 1977, Hammond 1976:78).

I have not been able to determine if these salinas were harvested during the Colonial period. Local informants recall that they were exploited as far back as the late nineteenth century. Gann (1918:22, see also Gann and Thompson 1931:236) reports that in the early nineteenth century in northern Belize salt was evaporated from sea water. Since I was unable to locate any other salinas in Belize, I assume he refers to Ambergris Cay. A more recent report by Lefond (1969:137) undoubtedly also refers to Ambergris.

THE SALINAS OF THE GULF COAST

The Gulf Coast region lies outside the Maya area, and thus beyond the focus of this study. Nonetheless, I decided to conduct a brief survey of the salinas of the region, for I found it somewhat puzzling that large quantities of Yucatecan salt were being exported to a region which, according to the literature, had a fair number of salinas. The following brief summary covers the known salinas of the coastal regions of Tabasco, Veracruz, and Tamaulipas (fig. 2.12); for a more detailed account, see Andrews (1980a).

Tabasco

Two salt sources have been reported in Tabasco, at Playa Jiménez (Zaráte 1917:29) and Iztapa (Mendizábal 1946a:297). Neither report was based on firsthand observation, and I was unable to locate salinas at these localities or any evidence of their past existence. Although Zaráte's report of salinas at Playa Jiménez is clearly based on a secondhand account, it nonetheless became embedded in the literature (cf. Mendizábal 1946a:297, Lozano García 1946:57, Lefond 1969:120). To the best of my knowledge, there are no other colonial or modern reports of salinas in the state of Tabasco.

Southern and Central Veracruz

The people of Veracruz exploited several small salinas in prehistoric and historic times, but they were few in number, and their production was minimal; none are in use today.

An early colonial account reports a small salt spring at Acalapa, 25 km east of the modern town at Minatitlán, where salt was gathered from a salt playa next to the spring. It was apparently not enough to supply local demand; according to the same account, the area was supplied with Yucatecan salt in the sixteenth century (Scholes and Warren 1965:784, Rathje 1972:325). This source was abandoned by the mid-nineteenth century (Williams 1852:230–31, Coe and Diehl 1980, I:17), and no trace of it remains today (Andrews 1980a).

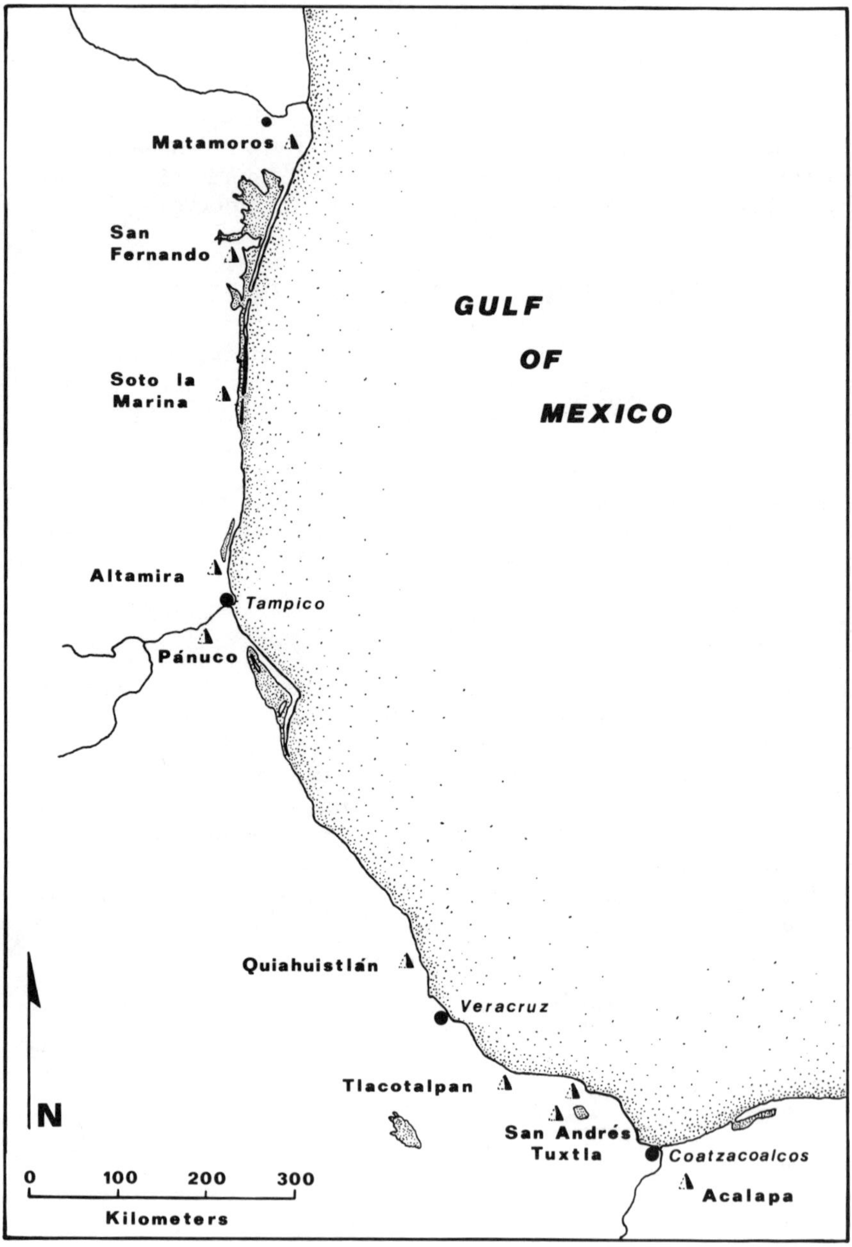

Fig. 2.12. Distribution of salt sources in the Gulf Coast region

Two sources have been reported from the vicinity of San Andrés Tuxtla. One of these, known as Las Salinas, is a small salt spring 8 km southwest of town, where salt was acquired by boiling brine. Luis Millet Cámara, who surveyed the area around the spring in 1978, found occupational remains dating to the Late Classic period, and some possibly dating to the Late Formative period. The salina was being exploited when the Spanish arrived in the sixteenth century (González de Cosío 1952:581–83), and continued to be until the early twentieth century (Zaráte 1917:29, Lozano García 1946:57, Medel y Alvarado 1963:358). Millet located another source, a small set of solar salt pans known as Salinas San Andrés, 33 km north of town, on the coast next to the cliffs of Peña Partida. According to local informants, these had once yielded small quantities of salt in the past, but it was bitter, and used only for salting fish. They were no longer worked in 1978. I found no reference to this source in the literature. The salinas of San Andrés were clearly too small for local needs; documents note that the area imported salt from Campeche when the Spanish arrived (Paso y Troncoso 1905–6, V:8, Mendizábal 1946a:285, Stark 1974:211).

Another minor source exploited at the time of the conquest was located at Tlacotalpan; again, this source yielded only small amounts, and the local population depended on Campeche for its salt needs (Paso y Troncoso 1905–6, V:3, Mendizábal 1946a:284–85, Stark 1974:209). I assume this salina was abandoned during the Colonial period, as there are no subsequent reports of its existence.

In 1976 I located some small salt playas next to two coastal lagoons at Quiahuistlán, 60 km north of Veracruz. These occasionally yield small amounts of salt in very dry years when the lagoons dry up, leaving a thin crust of salt that local farmers collected in the recent past. I was unable to find any documentation on this source, though it is possible that it may have supplied the nearby community of Quiahuistlán in prehispanic times.

In the early 1970s, a large salt refinery was established at Pajaritos, on the outskirts of Coatzalcoalcos. The salt, which is mined from a subterranean salt dome, supplies industrial salt to nearby oil refineries and table salt to the isthmian region and Chiapas. By the mid-1970s, production exceeded 150,000 tons. This industry is of no consequence to our reconstruction of the past salt industry, for deep salt domes were not exploited in prehistoric or historic times; moreover, there is no record of saltmaking in the Coatzacoalcos region prior to the 1970s.

Northern Veracruz and Tamaulipas

Several small lagoons between Tampico and Pánuco were exploited during the Colonial period, but appear to have been minor and unreliable sources of salt. The salt was acquired by solar evaporation, and production depended heavily on rainfall patterns: in very wet years none was produced. These salinas were located at Tampico, Chila, Pánuco, Tanchoy, and Cuymatlán (Mendizábal 1946a:286, Stresser-Pean 1971:589). It is clear that these sources were not capable of meeting local needs, for Yucatecan salt was imported into the Tampico-Pánuco area during the Colonial period and in the nineteenth century (Paso y Troncoso 1939–42 12:125, Thompson 1966:221, Regil and Peón 1853).

Along the coast of Tamaulipas are a number of estuarine lagoons that are well-suited for solar evaporation. Several of these are presently being exploited by modern industrial methods. Major salt works are found at Altamira (also known as Lomas del Real, or Salinas San Enrique), Soto la Marina, San Fernando, and Matamoros. These lagoons have been exploited since the middle of the eighteenth century when Spanish colonizers settled the area (Zaráte 1917:28 – 29, Mendizábal 1946a:302 – 3, Lozano García 1946:57, Lefond 1969:134, Martínez Vera 1979). I was unable to locate any evidence of saltmaking in this area prior to the eighteenth century; in prehispanic times this region lay beyond the frontier of the high culture areas of Mesoamerica and was inhabited by nomadic groups who probably had little need for salt. Even if the salinas had been exploited, their production would have been of little consequence to populations farther south.

It is apparent that the limited salt sources of the Gulf Coast region could not have satisfied even local demand. Clearly, Veracruz and Tabasco relied primarily on Yucatán for their salt needs. As a passage from the *Relación de Tlacotalpan* of 1580 notes, "Hacen sal poca y ruin; provehense della de Campeche . . . " (Paso y Troncoso 1905 – 6, V:3). There is little reason to doubt that this trade between Yucatán and the Gulf Coast had its origins long before the Spanish conquest. As Thompson (1953) has shown, substantial evidence indicates interaction between the two regions beginning in the Formative period. Whether the salt trade is of similar antiquity is a challenging question that awaits further research.

3.

The Salinas of Chiapas

THE HIGHLANDS

THERE ARE THREE SALT-MAKING LOCALITIES in the highlands of Chiapas: Ixtapa, Salinas Atzam, and La Concordia (also known as Custepeques) (fig. 3.1). The first two salinas are minor sources: Ixtapa supplied a small number of communities in the central highlands, and Atzam produces only minute quantities of salt used for ritual purposes. The third source, the salinas of La Concordia, was once a major source for the Central Depression of Chiapas. All three sources appear to have been exploited in prehispanic, colonial, and modern times, but little remains of the industry today. The salinas of La Concordia have been flooded by a modern dam, while the production of Ixtapa has declined dramatically in recent years.

La Concordia (Custepeques)

The salinas of La Concordia are located in the Central Depression of Chiapas, in the south-central region of Chiapas (fig. 3.1). Although these salinas were covered by the waters of the Angostura Dam, built in the early 1970s, several descriptions of them are available, the most informative being those of De la Peña (1951, IV:1206–09) and Blom (1945).

These salinas consisted of many brine springs that flowed into several rivers and springs in a 20-km radius of the now-flooded town of La Concordia. When the dam was built, the town was relocated on higher terrain and renamed Nueva Concordia. According to De La Peña, there were seven salt works exploiting twenty-seven salt springs. These were located along the Custepeques, Aguacate, Concordia, Agua Zarca (or Pasos Padres), and Morelia rivers, and along the Piquente, Malpaso, Arroyo, Grande, and Limonar streams.

La Concordia was the only place known in the Maya highlands where solar evaporation was used. This simple method involved tapping brine springs by means of large flumes made of tree trunks and wooden planks, and channeling their water into a series of small shallow evaporation pans built of stone and lime mortar (fig. 3.2). These pans irregularly shaped and of varying sizes, generally were

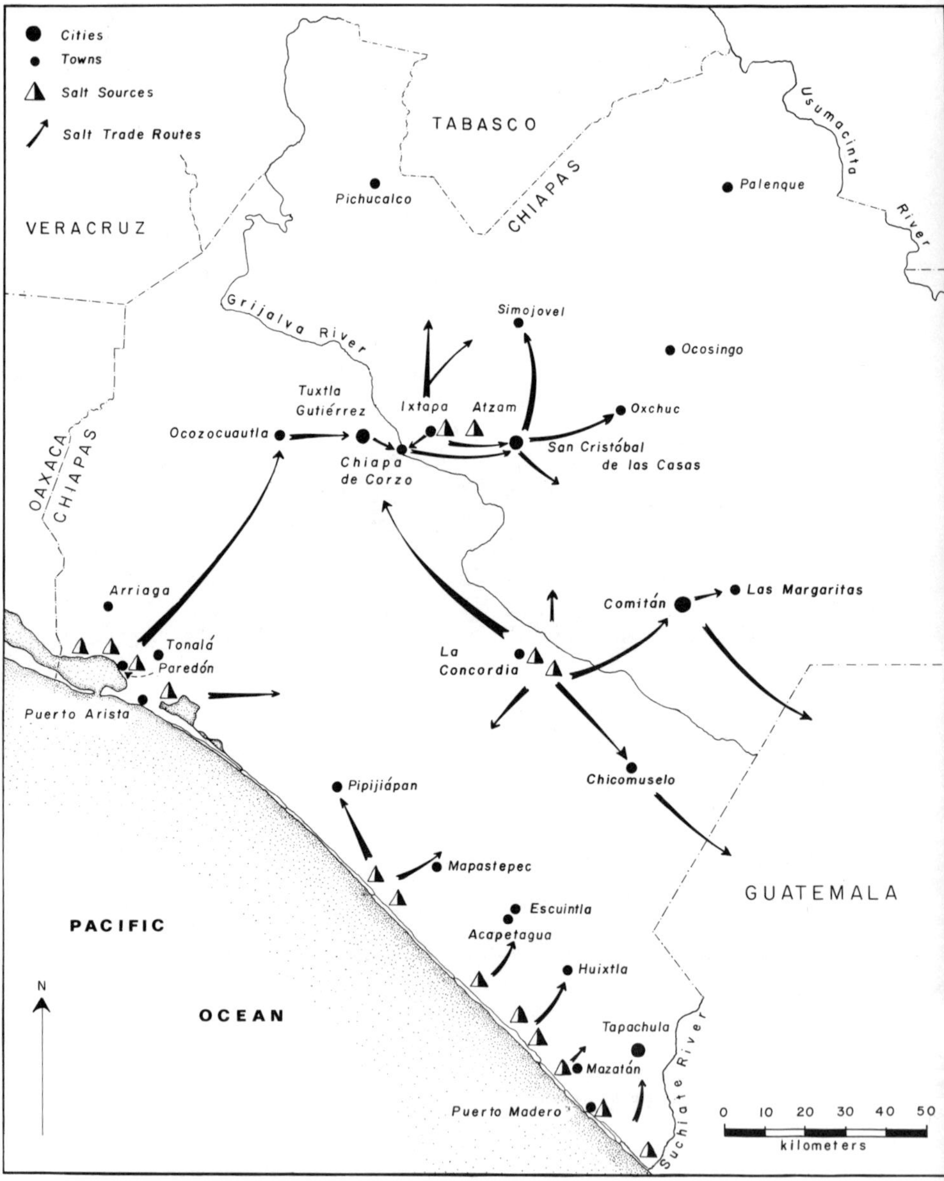

Fig. 3.1. State of Chiapas, showing major salt-making localities and principal past and present salt trade routes

Fig. 3.2. Solar evaporation pans at Salinas de Portatenco, La Concordia, Chiapas. (Photograph by Frans Blom, ca. 1945, courtesy of Gertrude Blom)

10 to 20 cm deep. Evaporation of the brine took place, on the average, in twenty to thirty days, after which the salt was raked and sent to market.

Saltmaking at La Concordia, as in other localities where solar evaporation was used, was a seasonal occupation, with work beginning at the onset of the dry season in January, and lasting until the first rains in May or June.

Another unique aspect of La Concordia was the manufacture of salt artifacts, a side activity for the saltworkers. These artifacts were made by placing variously shaped reed frames in the pans. Salt would eventually congeal around the frames, forming the desired objects (figs. 3.3 and 3.4). "Stars of David" were the most common artifact made, though five-pointed stars and Christian crosses were also manufactured. The only other known examples of the manufacture of salt artifacts in Mesoamerica are the mold made human figurines observed by Father Ponce in the 1580s at Atoyaque, Colima (Ponce 1873, II:120–21).

The salt artifacts of La Concordia, popular throughout the Central Depression and adjoining highlands, were bought by natives who placed them in churches as religious offerings (Eduardo Martínez personal communication 1977). One can only speculate if this

Fig. 3.3. Solar evaporation pans with stick frames for Star of David artifacts at Salinas de Portatenco, La Concordia, Chiapas. (Photograph by Frans Blom, ca. 1945, courtesy of Gertrude Blom)

practice also constituted a tribute to the church: it undoubtedly satisfied many a padre's need for salt! Blom suggests, tongue in cheek, that the Stars of David prove that the New World was originally populated by a lost tribe of Israel!

History. Lost tribes aside, there is good reason to believe that some of the salinas of La Concordia were exploited in prehispanic times. Explorations in the vicinity of the Salinas de Portatenco, undertaken by the New World Archaeological Foundation, have located a large prehispanic site on a spur directly above the salinas. While the architecture at the site is primarily Late Classic, ceramics from surface collections reflect an occupation span from Middle Formative through Early Postclassic times. Below the site, at the salinas proper, are several middens which also have Formative remains (Lowe 1959:49, personal communication 1977; Warren 1978).

Several other sites, spread along the north side of the Concordia River and located near salinas, have remains dating from Late Classic through Late Postclassic times (Warren 1978).

Subsequent to the prehispanic period, information on La Concordia is scant. The oldest account I have encountered is that of Pineda (1845:35–36), who observed that the brine had a red tint, which he attributed to an iron content. Later reports also mention the redness of the brine, and in some cases, of the salt itself. Pineda also noted that the salt was useful for salting meat. The next account is that of Sánchez (1915:161–62), who essentially repeats the same information.

Fig. 3.4. Star of David made of salt from Salinas de Portatenco, La Concordia, Chiapas. (Photograph by Frans Blom, ca. 1945, courtesy of Gertrude Blom)

Most subsequent accounts have little to say beyond reporting the existence of the salinas (Zaráte 1917:27; Mendizábal 1946a:298; Lozano García 1946:43; Mullerried 1957:144–45; Helbig 1964a:138–40, 1976, 2:142–43).

Production. De La Peña (1951, IV:1207) reports that the production of 1947 was 189 tons, that of 1948, 158 tons. His figures are probably reliable, as the salinas were incorporated into a *cooperative* which kept official records. He also notes that in good years the salinas yielded, at best, 200 tons, or enough to satisfy the basic needs of approximately 68,000 people.

Trade. From the above figures, it is apparent that the salinas of La Concordia could meet the demand of a relatively large region. Available records and reports from informants confirm this.

De La Peña (1951, IV:1208 – 09) reports that La Concordia salt was marketed throughout the Central Depression, in the municipalities of Comitán, Margaritas, Sapaluta, San Bartolomé, Pinola, Jaltenango, Socoltenango, and Chicomuselo. Helbig (1964a:140) reports that the salt was also sent to Tuxtla Gutiérrez, and Blom (1945:4) claims that some of it, known as "pink salt from Comitán," reached as far as Mexico City where it was sold as a remedy for rheumatism at $25 a kilo! According to a report written in 1883, salt from La Concordia was also traded into northwestern Guatemala in significant quantities during the nineteenth century (Manó 1940:511).

To conclude, then, the Salinas of La Concordia represent the largest salt source in the highlands of the Maya area, and the major source for a large part of the state of Chiapas. I gathered from informants that some of the Concordia salt even reached the western fringes of the lowlands via Las Margaritas. Even though the quantities were undoubtedly small, this is the only instance I know in which highland salt was traded to the lowlands.

Ixtapa

The salinas of Ixtapa lie at the bottom of the gorge of the Río Salinas, below the town of Ixtapa (fig. 3.5). They consist of a small brine well and seven long thatched huts, known as *cocinas* or cooking houses, where the brine is cooked. The well, carved out of bedrock to a depth of 2 m below the surface, taps an underground brine spring which flows into the Salinas River.

Probably more than 2,000 years old, the well is believed to be sacred. Directly behind it, on an ancient stone platform, are three large wooden crosses. Placing such crosses at sacred locations is a common pattern throughout the highlands of Chiapas and northwestern Guatemala (cf. Vogt 1969, Oakes 1951), and they are usually in groups of three. At the nearby salinas of Atzam, and at those of San Mateo Ixtatán in northwestern Guatemala, I also observed this pattern of placing three crosses next to "sacred" brine wells. Among the Tzotzil these crosses mark the location of a shrine which serves as a vehicle for communication with the ancestral gods and earth lords. Consequently, people often pray and burn incense in front of them. For some unknown reason, the crosses must occur in groups of three to be effective (Vogt 1969:387–90).

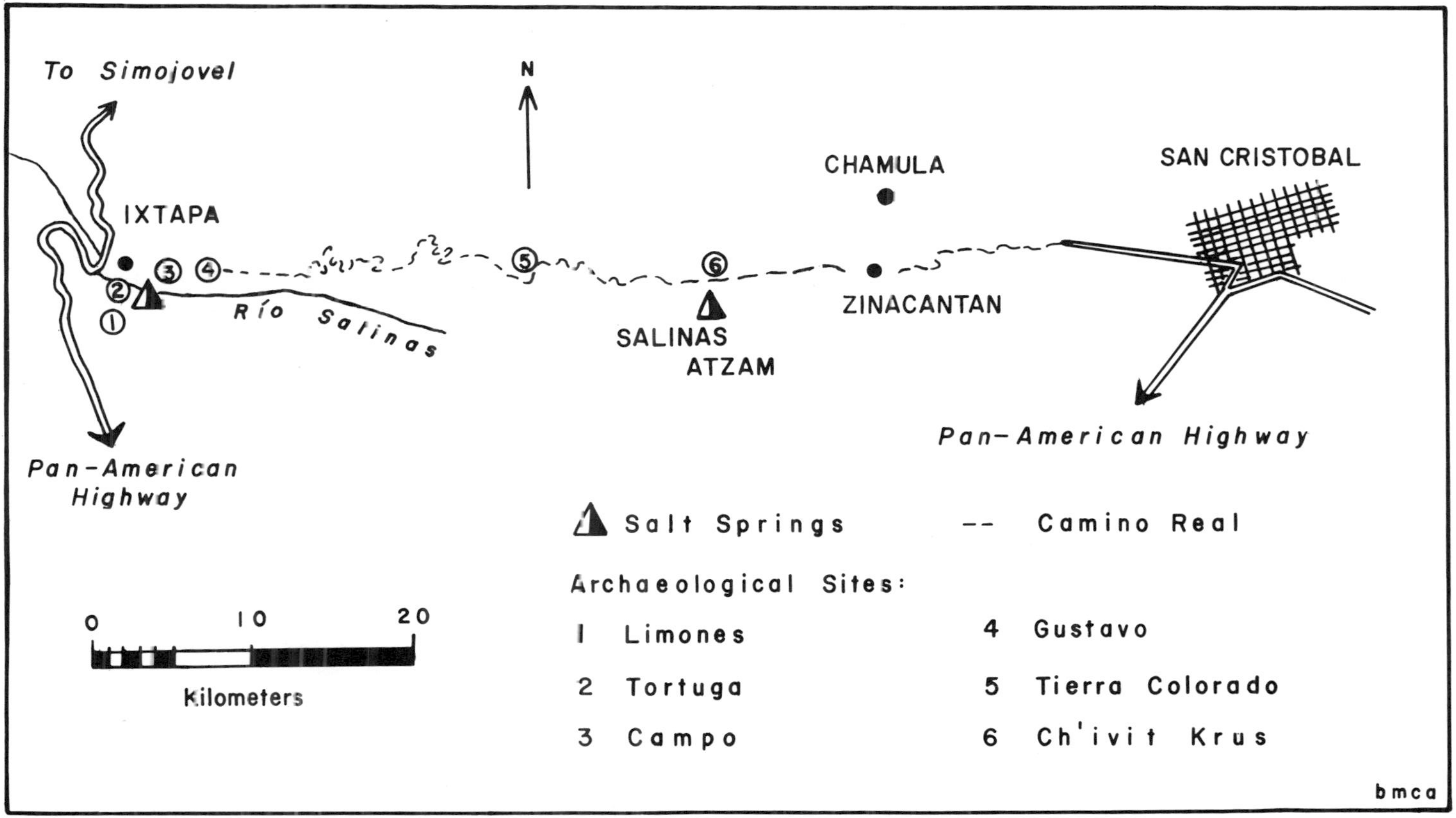

Fig. 3.5. Ixtapa–San Cristobal region, Chiapas. (Recrawn from McVicker 1974:549)

In the case of the sacred brine wells, the crosses are shrines dedicated to the ancestral gods, "owners" and guardians of the salinas. Local informants at all three salinas concurred that the purpose of the rituals is to keep those gods happy.

This ritual dimension of the salinas is important, as a number of Ixtapan families depend on the salt works for part of their livelihood. Tradition has it that seven families work the salinas where, since time immemorial, they have maintained exclusive control of the salinas by right of an ancient custom. I was unable to verify whether or not this tradition is still upheld; some of the older saltworkers maintain that it is, whereas shopkeepers in the village of Ixtapa noted that anyone who wished to exploit the well could do so. During the Colonial period the salinas were reportedly "communal" (Ximénez 1965, III:705–6 [Bk II, Ch. 48]), though as noted previously, the concept of communal ownership can be a legal fiction (cf. Zabala 1959).

In support of the seven-family tradition is the fact that there are seven cocinas (one had collapsed by 1976). The brine is cooked in these huts in *pangas* (large rectangular iron vats) atop woodburning adobe ovens. As the salt begins to congeal, it is packed into small cylindrical tubs made of *petate* (straw matting) (fig. 3.6), which are removed after the salt hardens. The resulting *benequenes,* or loaves, are then ready for sale. Each benequen weighs approximately 600 grams, although they were once larger, weighing as much as 1.5 kg (Helbig 1976:143).

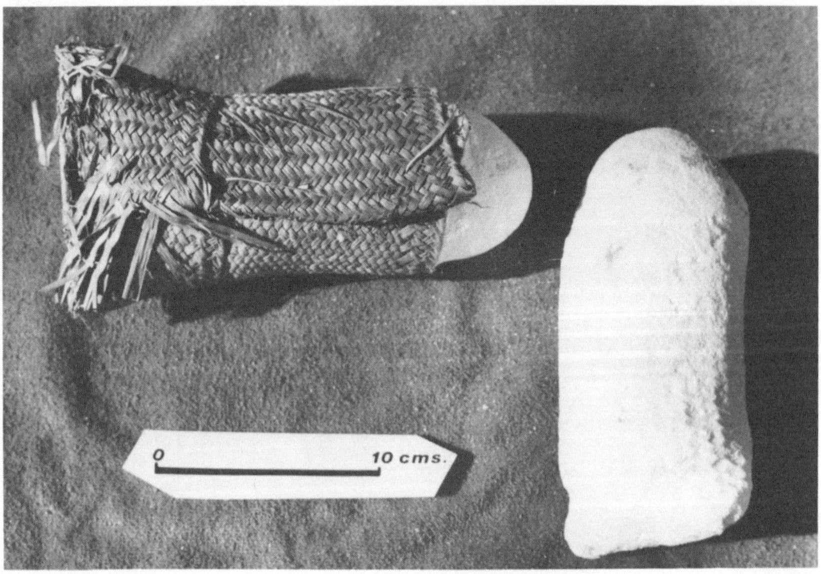

Fig. 3.6. Benequenes, or salt loaves, from Ixtapa, Chiapas

Several aspects of saltmaking in Ixtapa have changed in the last half century. Before 1950 the brine was transported in large, narrow-neck ceramic water jugs carried by a tumpline. These were rare in 1977. Although still in use in northwestern Guatemala, they have been replaced elsewhere by plastic containers and buckets (cf. Reina and Hill 1978:109).

Another feature that has changed is the cooking technology. The large iron vats have only been in use for approximately fifty years. Before iron sheeting was available, the brine was cooked in enormous earthenware vessels (approximately 90 cm deep and 70 cm in diameter). Although I was unable to locate any of these, I did observe several large, partially broken, narrow-neck jars still used for storing brine. According to the saltmakers, they are no longer manufactured, the potters who made them having died long ago.

The transition from ceramic vessels to iron vats appears to have been gradual, as the vessels were still being used for cooking brine as late as the 1940s (Gutiérrez H. 1949:26–27).

History. In the immediate vicinity of the salinas extensive archaeological remains have been surveyed and excavated by Donald E. McVicker (1969, 1978; see fig. 3.5). His work has revealed a long history of human occupation, from Late Preclassic times until the conquest. At the salinas proper, he found ceramic remains dating from Terminal Classic times through the Postclassic period and at the site of Campo de Aviación, directly above the salt well, he recovered materials dating to the Late Preclassic period. Consequently, he suggests that salt production began in Preclassic times (1969:276).

McVicker has identified a ceramic type as "bisquit ware" which may have been used in salt manufacture. Most bisquit ware vessels are large, flat-rimmed ollas that would have served perfectly as containers for cooking brine (cf. McVicker 1969:32, his fig. 28c). These vessels, common during the Late Classic period, are similar to brine-cooking vessels used along the Guatemalan and Salvadorean Pacific coast during prehispanic and colonial times.

That the salinas of Ixtapa were in operation at the time of conquest and thereafter throughout the Colonial period is well documented. I have located five reports that mention the salt works, dating to 1523 (Díaz del Castillo 1967:392–93 [Ch. 166]); 1591 (Navarrete 1966:99–103); 1691 (Remesal 1932, II:178); 1720 (Ximénez 1965, III: 705–6 [Bk II, Ch 48]); and 1774 (Feldman 1973:112).

According to Ximénez, in prehispanic times the salinas of Ixtapa were the cause of continual warfare between the Zinacanteco and Chiapanec peoples. Ixtapa was part of the Late Postclassic kingdom of Zinacantan, but because of its position near Chiapanec territory, was vulnerable to attacks from the Chiapanec, who were quite desperate to gain control of the salinas. Nonetheless, Zinacantan appears to have held on to the salinas; after all, the Zinacantecos were a nation of traders, and the salt trade was a fundamental part of their economic structure. Even after the conquest, in 1571, the Chiapanec attempted to lay legal claim to the Ixtapa salinas (Navarrete 1966:99–103). Apparently the attempt failed: the trade of Ixtapa salt is controlled to this day by merchants from Zinacantan.

During the late Colonial period and early nineteenth century, Ixtapa suffered a dramatic decrease in population. In 1774 the village had only forty inhabitants (Feldman 1973:113), possibly because local labor was being recruited into large agricultural plantations nearby (Wasserstrom 1976). As a result, the salt industry must have declined, but by the middle of the nineteenth century it had apparently revived (Pineda 1845:36; Wasserstrom 1976:488).

Subsequent references to the salinas are quite numerous (cf. Sapper 1897:188, 304; Sánchez 1915:161; Zaráte 1917:27; Blom 1932:536; Corzo 1943:83; Mendizábal 1946a:296; Mullerried 1957:145) and reveal the continuity of the industry. More important, however, are several twentieth-century ethnographic accounts of the salinas, particularly those of Gutiérrez H. (1949:25–27), De la Peña (1951, IV:1209–10), Zabala (1959), Helbig (1964a:92, 1976:143), Vogt (1969) and Wasserstrom (1970). The latter focuses on the complex social and ritual aspects of the salt industry and on the integration of Ixtapa with the regional network of Tzotzil towns.

Production. It is difficult to estimate the production of Ixtapa, as the saltmakers themselves have only a vague notion of how much they produce. They tend to think of production in terms of number of benequenes produced, rather than in terms of kilograms or tons. After a considerable amount of interviewing, followed by a lengthy analysis of tidbits from earlier accounts (cf. De la Peña 1951, IV:1210; Zabala 1959; Helbig 1976:143), I concluded that present-day production does not exceed 35 tons a year. Production, however, has declined: Ixtapa may have produced between 60 and 100 tons a year during the 1940s and 1950s. The figure of 100 tons is an ideal maximum, which could theoretically be attained only if all seven cocinas operated 365 days a year. This, however, is most unlikely, since the salt-making families must take time off to gather firewood and tend to their cornfields. Moreover, the rate of production depends on the availability of firewood, which is often scarce. Consequently, a family may stop making salt for several days, or in some cases, for several weeks.

Production has been declining for a number of reasons, such as inflation, rising labor costs, increasing scarcity of firewood, and the fact that many young people are leaving Ixtapa to find jobs elsewhere. The major factor, however, that has been undermining the industry since the late 1950s is competition from industrial salt. Improved transportation networks have been supplying highland markets with cheap table salt, most of which comes from Tehuantepec or Coatzacoalcos. However, because of its purported medicinal qualities (cf. Pineda 1845:36) and unique taste, Ixtapa salt retains a small niche in highland markets. It is still available every day in the market at San Cristóbal de las Casas, where it is sold at a slightly higher price than industrial table salt.

Trade. The salt trade of Ixtapa has traditionally been, and still is, controlled by merchants from Zinacantan who come to Ixtapa to buy the salt, and then sell it in communities throughout the central highlands. Local informants told me that in the past an occasional shipment went to Chiapa de Corzo, to the southwest; most salt, however, has traditionally been traded to the north and east.

A large part of the salt is traded to nearby Tzotzil communities, such as Zinacantan and Chamula, as well as the modern city of San Cristóbal de Las Casas, a major market for the salt of Ixtapa. People from many surrounding communities come to the marketplace at San Cristóbal where they purchase the salt in both retail and wholesale quantities.

From San Cristóbal, the salt is transported east. Villa Rojas (1946:554) reports that in 1944 Ixtapa salt reached as far as Oxchuc at the eastern edge of the highlands. It was also taken to Amatenango (Siverts 1969).

From Ixtapa itself, a fair amount of salt went north, as far as the communities of Soyalo, Bochil, Jitotol, El Bosque, and Simojovel (Zabala 1959; McVicker 1969). Simojovel was famous in prehispanic and colonial times for its amber mines, whose trade was controlled by Tzotzil and Zoque merchants (Navarrete and Lee 1969). We can infer that Zinacanteco (Tzotzil) merchants were probably supplying Ixtapa salt in exchange for Simojovel amber (McVicker 1969:261).

Available data indicate that the Ixtapa salt trade had a limited regional range—to the north and east—of less than 50 km. Such a range fits well with its production capability. If Ixtapa ever produced its maximum potential of 100 tons a year, that amount would only satisfy the basic dietary needs of less than 35,000 people. In 1960 the Tzotzil numbered well over 100,000 people.

Salinas Atzam

Salinas Atzam is a tiny source of salt whose annual yield is only a few hundred kilograms. Like Ixtapa, the source of salt is a small well that taps an underground salt spring. The well water dries up during the dry season, and the brine is taken to houses where it is cooked in large earthenware ollas similar to those used in Ixtapa for storing brine. The final product is a coarse-grain salt.

The small quantities of salt produced are consumed locally and by other nearby hamlets. The importance of the salt rests with its purported sacred and medicinal qualities, and it is valued in the neighboring region for use in ritual and healing ceremonies. For this reason the salt-making activities survive virtually unchanged since prehispanic times. Wasserstrom (1970) and other members of the Harvard Chiapas Project (cf. Vogt 1969) have studied the numerous and complex ritual and medicinal aspects of Atzam salt.

Salinas Atzam is a small hamlet of approximately twenty houses, a school, and a refurbished church. The salt well is directly behind the church under the shadow of three large wooden crosses. The curbstone of the well also has a Christian cross on it, made by inset plain white tiles. A strong religious aura pervades this area, which is the site of an ancient shrine. The church is dedicated to the Virgen del Rosario, the guardian of the salt well.

Atzam lies in the municipality of Zinacantan and has important ritual ties to the head town. One of its main functions is to furnish the elders of Zinacantan a regular supply of salt for their use in ritual affairs. Cancian (1965:36–37) believes this custom may be a vestige of an ancient tribute paid by Salinas to Zinacantan.

History. The history of Atzam began sometime in the prehispanic period, as the extensive archaeological materials in the vicinity clearly indicate. The hamlet itself has numerous middens, in which I located at least one diagnostic sherd: Balancan Orange, a type of pottery which dates to the Terminal Classic period. I believe that Atzam was occupied at this time, because of the presence of a large archaeological site of similar date on a ridge overlooking the modern hamlet. This site, known as Civit Krus, has remains dating to Terminal Classic and Early Postclassic times (McVicker n.d.).

I have been unable to locate any colonial references to Salinas Atzam, but the hamlet was occupied during the Colonial period. It was reportedly abandoned at the end of the eighteenth century and reoccupied in the mid-nineteenth century (Wasserstrom 1976:487–88).

Later reports on Salinas Atzam include those of Zaráte (1917:27), Mendizábal (1946a:296), and Mullerried (1957:144–45), all of whom allude to salt works as being the salinas of Zinacantan.

THE PACIFIC COAST

Few traces remain of the once active salt industry of the Pacific coast of Chiapas, which once extended from the Oaxaca border, near the isthmus of Tehuantepec, to the Suchiate River at the Guatemalan border. As of 1978 the only remaining salt works were a few small operations in the vicinity of Tonalá and Arriaga, on the northwest coast (fig. 3.1).

Salt-making Technology

Sal Cocida. Making salt by the sal cocida method was common all along the coast from Mapastepec to Puerto Madero. Though long abandoned, traces of these salt works are evident in the form of large mounds of earth which were the byproduct of the sal cocida process. These "salt-making mounds," which in some cases are as high as 4 m, are often confused with prehispanic structures. They are found near Mapastepec, Escuintla, Acapetagua, Mazatán, Huixtla, and Puerto Madero. At several of these localities, sal cocida operations were still being carried out in the early 1960s.

Tapesco Method. Another method for making salt, unique to the Chiapas coast, is the tapesco system (Corzo 1943:98–99; De la Peña 1951, IV:1204–5). Simply put, it involves leaching marsh soil with estuary water which is then evaporated in small solar pans (figs. 3.7 and 3.8). Like sal cocida, the tapesco salt-making operation is seasonal.

The soil to be leached is raked up. During the dry season from January to May or June, when the estuaries have receded, it is placed in raised wooden trays known as *tapescos* (fig. 3.7). The bottom part of these have a three-layered filter made of thin sticks covered with grass reeds, which in turn are covered by a layer of sand. Salt water is drawn from small shallow wells that tap the brackish water table. It is then poured over the soil in the tapesco, thus leaching the soil. After passing through the filter it collects in a mud-lined pan beneath the tapesco, from which it is drained into a storage tank adjoining the tapesco. The sunken storage tank, whose walls are made of plaster-coated mud, is known as a *taza* (fig.3.7).

The brine is then placed in a series of small shallow solar evaporation pans, known as *pilas*, whose floors and walls are also made of plaster-coated mud. These pans measure, on the average, 150 by 150 cm and are 10 to 15 cm deep. They are arranged in rectangular grids—two rows in each grid—and each grid has between ten and eighteen pilas (fig 3.8).

Depending on available sunlight, evaporation will take from four to six days, after which time the salt is raked into piles with wooden hoes.

The sal cocida operations represent an ancient industry along the Pacific coast of the Maya area, where it has been traced to Formative times. The antiquity of the tapesco method, however, is not clear. Unlike the sal cocida operations, the tapesco salt works leave few, if any, traces. The wooden frames of the tapescos are perishable, and the mud walls of the tapescos, tazas, and pilas wash away in one or two rainy seasons. In fact, many must be rebuilt every year.

Salt-making Localities

It is difficult to pinpoint specific salt-making localities along the Pacific coast of Chiapas, since most areas have at one time or another been the site of such activities. On the basis of limited available data, I have arbitrarily isolated the salinas into two major regions: the Tonalá/Arriaga region and the Soconusco coast.

Fig. 3.7. Tapescos and storage tank, or taza, at Salinas de Punta Flor, Arriaga, Chiapas. Note drains leading from beneath the tapescos to the taza.

Tonalá/Arriaga Region. This area traditionally has had the largest salt works of the Chiapas coast. These salinas were located around two lagoons, the Mar Muerto and the Laguna La Joya. The former included the salinas of Berlín, Moctezuma, Paredón, Punta Flor, and La Gloria; the latter included Capulín, Cabeza del Toro, and Puerto Arista. In 1978 only the salt works of Berlín, Punta Flor, and Cabeza del Toro were in operation, and the saltworkers there did not expect the industry to survive many more years. They employ the tapesco method, which is labor-intensive, and cannot compete with industrial salt from Tehuantepec and Coatzacoalcos. Production has declined, and what is produced is intended primarily for cattle ranches in the region. I was unable to locate any traces of a sal cocida industry in this area. Local informants claimed that the tapesco method had always been the traditional mode of making salt.

The salinas of this area, often referred to as the salinas of Tonalá, are well documented in the literature of the region. Detailed accounts of the salinas include those of Corzo (1943:98–99) and De la Peña (1951, IV:1204–5). Briefer references include those of Pineda (1845:35), Zaráte (1917:27), Mendizábal (1946a:285), Lozano García (1946:43), Helbig (1964b:78, 1976, 2:142) and Lefond (1969:132–33).

Fig. 3.8. Pilas, or solar evaporation pans, with recent harvest of salt at Salinas de Punta Flor, Arriaga, Chiapas

The Soconusco Coast. Very little is known of the now-abandoned salt works of the Soconusco coast, which employed both the sal cocida and tapesco methods. These salinas were located near the coast in the vicinity of Mapastepec, Escuintla/Acapetagua, Hiuxtla, Mazatán, and Puerto Madero and the Suchiate River.

Carlos Navarrete, who has conducted a detailed archaeological survey of the Soconusco coast, informs me that sal cocida mounds are found at all of these localities, and that the tapesco method was employed until recently near Mazatán and the mouth of the Suchiato River (personal communication 1978). Local informants also reported recent tapesco operations in the Mapastepec and Escuintla areas.

Literature references to the salinas of Soconusco are few, and none provides any details regarding production. The earliest report is that of Juarros (1823:20, 1936:18), which makes general reference to the salinas of the region, and particular mention of some salinas at Hacienda San Pablo, a site I was unable to locate. Later, in the mid-nineteenth century, Pineda (1845:36) reports that the Soconusco coast had nineteen salt works, but he does not elaborate.

Twentieth-century reports are equally uninformative. Most sources report the existence of salinas at Guayatengo (or Huayalotengo) near Mapastepec, and at Guamuchil (or Huamuchil) near Escuintla, but do not describe them (Zaráte 1917:27; Mendizábal 1946a:285; Lozano García 1946:43, De la Peña 1951, IV:1206; Helbig 1976, 2:142; Lefond 1969:132–33). Lefond lists four other salinas in the Soconusco region, but I was unable to locate any of these: Punta de Agua, San Ramón, Laguna Redonda, and El Amparo.

Prehispanic History

The Pacific coast of Chiapas may have one of the earliest salt-making sites in Mesoamerica. Excavations at the site of El Pajon, supervised by Maricruz Paillés H. (1978:88–89), have revealed "areas of burned earth, possible hearths or ovens and great numbers of sherds from burned *tecomates* (neckless, globular vessels)." These remains date to the Early Dunas phase (1000 to 900 B.C.), and Paillés believes they are the remnants of a very early sal cocida operation. While the evidence is by no means conclusive, her interpretation of the data seems quite reasonable.

Elsewhere along the coast there is only indirect evidence of saltmaking in Formative times. One likely area of production is the Mazatán area, which includes the Early Formative site of Altamira (Lowe 1975) and several other sites of Early and Middle Formative date (Navarrete 1974). Another site nearby, known as Alvarez, also has Formative remains, as well as indications of recent salt-making activities (Lowe personal communication 1976). It is likely that these sites were involved in saltmaking in Formative times.

Evidence of saltmaking in subsequent periods is scant. Navarrete (personal communications 1978) has located a number of Classic and Postclassic sites along the coast, several of which appear to be associated with prehispanic salt works. His forthcoming report will contain descriptions of these sites.

One site that was undoubtedly involved in saltmaking is the Postclassic site of Paredón, near Tonalá (Drucker 1948:168, Navarrete 1959). This site lies in the midst of the largest salt-working area on the coast, and future excavations there will probably reveal evidence of saltmaking. Finally, Drucker (1948:158) reports two Postclassic sites on the southeast coast, near the mouth of the Cahuacán River (between Puerto Madero and the Suchiate River), known as El Chical and El Jato. Given their estuarine location, it is quite possible that the inhabitants of these sites also made salt.

Colonial and Modern History

I was unable to locate a single reference to saltmaking on the Pacific coast prior to the nineteenth century. This is unfortunate, as it leaves us in total ignorance about the industry at the time of conquest or, for that matter, throughout the Colonial period. This lack of any information leads me to believe that saltmaking was a marginal activity during that period.

Production and Trade

Salt production along the coast of Chiapas appears to have been a fluctuating cottage industry without any central control or record-keeping. Consequently, it is impossible to arrive at any accurate estimates of past production.

The only available data come from the Tonalá area, which has always been the largest producer. Forty years ago, Corzo (1943:98–99) estimated that the annual production of the Tonalá area was 1,000 tons. The industry has since declined considerably. In 1978 I made a survey of the area and arrived at an estimate of 300 tons a year.

Production on the Soconusco coast was undoubtedly much lower, as the large number of freshwater rivers draining into the estuaries seriously hinders the buildup of salt deposits in the estuarine soils (De la Peña 1951, IV:1211).

As an educated guess, I would estimate that the production of the entire coast of Chiapas has never exceeded 2,000 tons a year. Such an amount would have served regional purposes, as it could satisfy the minimal dietary needs of 685,000 people. According to informants throughout the state of Chiapas, the trade of Pacific coast salt was restricted to the coastal plain, central highlands, and Central Depression. The salt from Tonalá was traded the farthest, to Ocozocuatla, Tuxtla Gutiérrez, and San Cristóbal de las Casas.

In 1978, the industry was moribund, and the remaining salt works at Tonalá supplied only a limited neighboring area. Chiapas now depends on salt from Tehuantepec and Coatzacoalcos.

4. The Salinas of Guatemala

THE PACIFIC COAST

THE COASTAL SALT INDUSTRY OF GUATEMALA spans the entire Pacific Coast from the Suchiate River on the Mexican border to the border of El Salvador (fig. 4.1). To arrive at an idea of its extent one need only consult the topographic sheets of the Instituto Geográfico Nacional: several hundred salinas are located along the shores of the Pacific.

Since this industry is so extensive, my survey had to be selective. While I did attempt to visit numerous localities, I sought different kinds of salinas. In addition, an aerial survey of the coast enabled me to arrive at a broader perspective of the distribution of the salt works. The literature on this industry is voluminous, and each major salt-making area has been well documented from the time of the Spanish conquest on (cf. Andrews 1980a: Appendix B).

Salt-making Technology

Before the twentieth century, all of the salt produced along the Pacific coast of Guatemala was made by means of the sal cocida process, which has been in use for more than 2,000 years.

When the Spanish arrived in Guatemala in the early sixteenth century, they encountered a widespread sal cocida industry, which several chroniclers described in detail. The best accounts are those of García de Palacio (1927:72), Estrada and Niebla (1955:79), and Fuentes y Guzmán (1932:33, 2:102–3). They all concur on the details of the salt-making process, which involved placing marsh soil in large canoes, filtering estuary water through the soil, and then cooking the resulting brine in large ceramic ollas.

Ollas were used until the late nineteenth century, when they were replaced by large iron vessels known as *peroles* (fig. 4.2). I was unable to locate any of the old ceramic ollas but did find fragments of them at several places along the coast of Guatemala and El Salvador. Judging from the remains, the ollas varied in size: depth ranged from 60 to 100 cm, while rim diameters varied between 60 and 80 cm; the thickness of the rim sherds was 2 to 2.5 cm, whereas the bottom sherds ranged between 4 and 5 cm. These sherds were

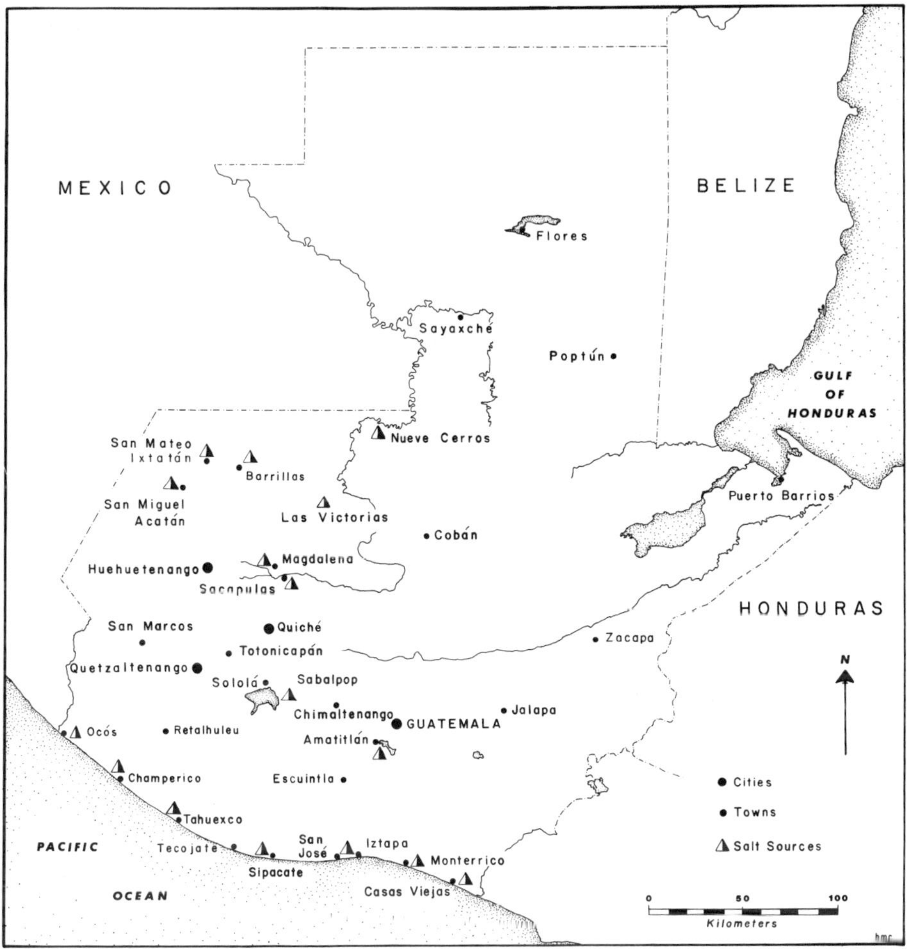

Fig. 4.1. Location of major cities, towns, and salinas in Guatemala

located in Colonial period and nineteenth-century middens, but they might as well have been prehispanic, since they are identical to shords collected by Edwin Shook (personal communication 1977) at the Late Formative site of Salinas Acapán, near Champerico. They are also similar to sherds of prehispanic salt-making vessels reported from Ocós (Coe and Flannery 1967:pl. 28k) and from San José (Shook 1949:15–16, fig. 46). The Ocós sherds, which date to Late Formative times, attest to a remarkable continuity in vessel form for almost 2,000 years.

It is not clear when iron peroles began to replace the ceramic ollas; informants tell me they date to the beginning of the nineteenth century, but ceramic ollas were still in use until the 1880s. On a visit to the Champerico area in 1882, Otto Stoll (1886:174) observed that both clay *and* "copper" ollas were employed for cooking brine.

Fig. 4.2. A perol, or iron vessel, used along the Pacific coast of Guatemala and El Salvador in the nineteenth and early twentieth centuries for cooking brine. (Tape measures one meter.)

The use of iron vessels did not last long in most places, as they were eventually replaced by the more efficient iron vats placed over adobe ovens, known as *hornos*. Informants tell me that these were introduced in the late nineteenth century; they were also in use at the highland salt source of Salinas Magdalena in the early 1880s (Manó 1940). All contemporary sal cocida operations use the hornos, and in some localities saltmakers still leach soil in large, ancient canoes. At one time, most operations employed two canoes, one placed atop the other. Soil rich in salts was placed in the upper canoes over a mat of reeds that served as a filter. The bottom of this canoe, which was drilled full of holes, acted as a colander, and brine which was obtained by leaching the salt from the soil in the upper canoe collected in the lower one (cf. McBryde 1947:pl. ld). Now they employ a single canoe with a double bottom and drain the brine by means of flumes made of palm trunks into storage tanks in the kitchens.

The main reason for the decline of the sal cocida industry is the development of solar evaporation technology. This process had its beginnings in the early twentieth century, when saltworkers dammed parts of estuaries, creating solar evaporation tanks known as *asoleaderos*. During the dry season evaporation leaves a salty brine, which is then cooked in the hornos. This method, a combination of solar evaporation and cooking, has obvious advantages, the most important being elimination of the soil-leaching process. The use of asoleaderos and hornos is still somewhat common in Guatemala and El Salvador.

The most important innovation in the Guatemalan salt industry was the introduction of modern solar evaporation during the 1920s. This process involved the construction of large asoleaderos and extensive patios of small shallow evaporation pans made of tiles. Although most operators now use tile, in some cases the pans were made of wood or concrete.

The production process is simple: brine from the asoleaderos is pumped into the shallow pans, where evaporation takes place in a day or two. The salt is then raked up, ground up, and iodized, resulting in a product known as "sal solar."

The sal solar industry developed slowly and did not become the dominant mode of production until the 1950s. By the 1980s solar operations were located all along the coast, the largest found near the ports of San José, Tahuexco, Sipacate, and Champerico.

Salt-making Localities and Available Data

Most of the salinas of the Pacific shoreline are clustered in seven regions:

Ocós. The area around Ocós has a number of sal cocida salinas and one sal solar operation. The industry here, although small, is very old, dating back to the Late Formative period (Coe and Flannery 1967:91–92). Unfortunately, this is also the least documented locality. I found no colonial records of saltmaking at Ocós, and only one reference to it in the mid-nineteenth century (cf. Coe 1961:19). Modern accounts of the industry also are lacking.

Champerico. This area, a long-time major producer of salt, is the second largest in the country. As at Ocós, there is clear evidence of saltmaking in Late Formative times at Salinas Acapán and at Salinas de Ixtán (Kidder 1940, Coe 1961:148–49, Shook personal communication 1977). There are numerous colonial references to the industry of this region dating from 1548, as well as several nineteenth-century accounts.

There are several salinas in the vicinity of Champerico today, the largest being the sal solar operations of Salinas Chapán, Ixtán, and Acapán.

Tahuexco. This region has many salinas which employ both sal cocida and sal solar methods. There are no reports of prehispanic sites—the area has yet to be surveyed—though the Spanish encountered a native industry here in the sixteenth century. Colonial period reports date from 1579 on, and there are several nineteenth-century accounts.

The salt industry of Tahuexco was studied by McBryde in the 1940s. At the time, the *salineros* were making salt by the traditional sal cocida method.

Sipacate. Both sal cocida and sal solar were produced in the Sipacate region in the 1970s. As in the Tahuexco area, no archaeological sites have been reported from the vicinity of Sipacate, but colonial records indicate that salt was being produced there in 1548. Several other sources make reference to the salt industry during the Colonial period.

San José/Iztapa. This area has long been the largest producer of salt in Guatemala. The first sal solar operations were established in the 1920s, and the industry has expanded enormously. More than half the salt produced in Guatemala comes from this region (Félix Montes personal communication 1978). The largest operations are those of San José, Santa Rosa, and Santa Marta, all located east of the port of San José (fig. 4.3). A few smaller operations are located near Iztapa.

Saltmaking in this area began in prehispanic times, evidenced by several archaeological sites, whose occupation spans 2,000 years

Fig. 4.3. Aerial view of Salinas San José, Guatemala. This modern solar operation was one of the largest in Guatemala in the 1970s.

from Late Formative through Colonial times; of these, two are located adjacent to modern salt works, and it is most likely that their inhabitants were engaged in saltmaking (Shook 1949, 1965; Walters 1977).

The industry is well documented from the mid-sixteenth through the nineteenth centuries.

La Avellana/Monterrico. This area is not well known archaeologically or historically. The oldest reference to saltmaking dates to 1601, and subsequent reports only contain passing references to the industry. In the 1970s there was a cluster of sal cocida operations east of the coastal hamlet of Monterrico.

El Ahumado/Casas Viejas/Garita Chapina. This region is also poorly known. No archaeological sites have been reported, and Colonial period references are scarce. Nonetheless, the fact that two nearby towns paid a salt tribute in 1548 would indicate that a native industry existed at the time of conquest (AGI Guatemala 128).

Approximately a dozen sal cocida operations now exist in the area, at least two of which were switching to solar evaporation in 1977.

Historical Summary

The salt industry of the Pacific coast of Guatemala began some time during the Formative period and was definitely underway 2,000 years ago. Archaeological data from Ocós and Champerico clearly suggest saltmaking activities at that time, and while the evidence from San José is less conclusive, it, too, was most likely an early salt-producing area. Evidence from other localities is not available simply because they have yet to be surveyed; nonetheless, I believe that future archaeological excavations will reveal evidence of prehispanic saltmaking at all the localities where a modern industry exists. Colonial records attest to the existence of salinas at all of the modern localities, except at Ocós, when the Spanish arrived in the sixteenth century. Moreover, the best accounts make it clear that these were *native* operations, whose mode of exploitation was something of a novelty to the Spanish (cf. García de Palacio 1927:72, Estrada and Niebla 1955:79).

The methods of the natives continued to be used throughout the Colonial period, even at salinas where the Spanish took control. The Spanish did, in fact, take over many salinas, which they exploited by means of a brutal system of conscripted labor (Cortés y Larraz 1958, II:223). The *Archivo General de Centro America* (AGCA) in Guatemala City, contains numerous documents that record complaints by native communities pressed into working at salinas (cf. AGCA 1689, 1690); in a few cases, they were released from service (AGCA 1672).

Confusion—and a fair amount of litigation—resulted over who had the legal right to exploit the salinas during the Colonial period.

Royal authorities did not follow consistent policy either. To promote the growth of the salt industry, the king decreed in 1573 that the saltworking localities be divided among Spanish landowners. Later, however, in 1591 and 1730, royal decrees reaffirmed the right of native communities to exploit their own salinas (AGCA 1573, 1591a, 1730). Yet we know from the account of Cortés y Larraz (1958, II:223), written in 1770, that the Spanish continued to control many salinas.

As in Yucatán, the crown took an early interest in the salinas, which were perceived as a source of tax income. As early as 1553 the king was inquiring about the income generated by the saltworks. In 1572 he requested to know whether the industry might be better served in the hands of Spaniards (AGCA 1553, 1572), and in a 1573 decree granted the salinas to them.

The royal decrees of 1591 also affected Guatemala, but as in Yucatán, they were not enacted until 1603. These decrees apparently had limited success and were rescinded in 1610 (AGCA 1591a, 1591b, 1610). Still, salt continued to be taxed throughout the Colonial period, and occasional complaints were made about excessive levies (cf. Fuentes y Guzmán 1932–33, I:150). Yet another parallel with Yucatán was the use to which the Guatemalan salt taxes were put. By the late sixteenth century, pirates had moved into the Pacific, attacking Spanish ships and ports. Most notorious among them was Sir Francis Drake. To counter this threat King Phillip II deployed in the Pacific an armada, known as the Flota de Barlovento, with Guatemalan salt revenues earmarked for the fleet's maintenance (AGCA 1591b).

Despite these various legal complications, the salt industry remained relatively stable until the late nineteenth century, when technological innovation began to affect production. The introduction of peroles and hornos increased production to a certain degree. As previously noted, a major change came with the introduction of asoleaderos, which eliminated the soil-leaching process, reduced the amount of labor involved in saltmaking, and brought about a pronounced increase in production.

The final innovation was the introduction of solar technology. The first sal solar operation was established at San José in the 1920s; others followed in the Champerico region in the 1940s and 1950s and spread along the coast. In the late 1970s, solar salt works accounted for 75 to 80 percent of the total national production (Félix Montes personal communication 1978).

The sal cocida industry will probably soon disappear, for several reasons. One is the growing scarcity of firewood, used in large quantities in the sal cocida process. In addition, development of large agricultural and ranching complexes has led to deforestation of huge tracts of jungle along the coastal plains of Guatemala and El Salvador. As a result, many sal cocida operations have ceased.

Those that survive depend on dwindling stocks of mangrove wood along nearby estuaries.

An even more important factor in the decline of the sal cocida industry is its economic feasibility. It is labor-intensive, and compared to solar operations, low in yield. In short, sal cocida cannot compete with solar salt. Most contemporary sal cocida operations are small affairs which produce crude, unrefined salt for small communities and cattle ranches. They survive only because Guatemala is not self-sufficient in salt and because saltmaking is a seasonal activity that provides income during the year when there is little agricultural activity.

Production

One of the more frustrating aspects of my research on the Pacific coast industry was my inability to locate any pre-twentieth-century production figures. I suspect that there is information on sixteenth-century production in one of the archives in Spain, as royal decrees requested reports on the productivity of the industry in 1553 and 1572 (AGCA 1553, 1572). Future archival work will likely turn up the responses to these royal requests.

Figures are available, however, for the twentieth century. According to Félix Montes (personal communication 1978), one of the leading producers of coastal salt, production was relatively low prior to the introduction of solar evaporation. He estimates that the entire coast produced less than 8,000 tons a year before 1920, when sal cocida was made with iron vat hornos or peroles, and asoleaderos. Before that period, when all sal cocida brine was manufactured from leached soils and cooked in ceramic ollas, production would have been considerably lower. My own observations lead me to estimate that production in prehispanic and colonial times was approximately 5,000 tons a year.

The introduction of sal solar operations did eventually increase production, but only gradually. In 1941 production was only 9,800 tons, and two years later 12,500 tons (Consorcio Salinero de Guatemala 1942–45, Anonymous 1944).

Even after 1950, when solar evaporation had spread to several localities, production increased slowly. One set of unpublished figures shows that between 1950 and 1972 average annual production was approximately 14,000 tons (RG, DGE n.d.). Most salt industry officials, however, commented that such a figure was too low: 20,000 tons a year appears to be a more reliable estimate. In 1978 production was approximately 30,000 tons, almost four times the level of the 1920s.

Since the late nineteenth century, Guatemala has imported salt from Chiapas, Belize, England, the United States, Germany, and El Salvador (RG,MHYCP, *M* 1880–1900; Manó 1940:511; Stoll 1886:172–73). Prior to 1960, these imports ranged from 84 to 3,000 tons

(RD,MHYCP, *M* 1880–1955; Zaráte 1917:4). After 1960 most of the imported salt came from El Salvador, and in 1976 Guatemala imported 20,000 tons from its neighbor (DGEYC, *AE* 1976).

Trade

Salt from the Pacific coast of Guatemala has traditionally been marketed in nearby communities of the coastal plain and adjoining highlands. Each salt-making locality supplied its own adjacent highland area. A general reconstruction of the trade range of each region can be arrived at through tribute lists, historical accounts, and information from informants. A valuable contribution comes from Feldman (1971), who presents a detailed reconstruction of trading patterns of southeastern Guatemala.

To judge from the available information, the salt trading patterns changed very little between prehispanic times and the early twentieth century, when the larger producers began shipping salt into the interior to Alta Verapaz and Petén. Briefly, then, the trade has been along the following lines (cf. Andrews 1980a for relevant references):

Ocós Region. Ocós is a minor producer and supplies only a small area in its environs. I found no evidence that Ocós salt has ever been traded to the highlands.

Champerico Region. This area has traditionally supplied the southwestern corner of Guatemala, namely the area comprising the present-day departments of Retalhuleu, Quetzaltenango, San Marcos, and parts of Totonicapán and Huehuetenango.

Tahuexco Region. Tahuexco salt has traditionally been marketed in the area of the present departments of Suchitepequez, Sololá, Totonicapán, and southern Quiché.

Sipacate Region. This region has been the traditional source for parts of the departments of Suchitepequez, Escuintla, and Chimaltenango.

San José/Iztapa Region. The salinas in this region probably were the main source of salt for the Valley of Guatemala and surrounding areas in prehispanic and colonial times. This market area includes the present-day departments of Escuintla, Sacatepequez, Chimaltenango, Guatemala, and Baja Verapaz. Feldman's study includes a detailed reconstruction of the protohistoric trade networks of the area, and coastal salt undoubtedly was transported over many of the routes he describes (Feldman 1971:figs. 4 and 6).

La Avellana/Monterrico Area. Known in ancient times as the coast of Guazacapan or Chiquimuilla, this region supplied the modern departments of Santa Rosa and Jalapa.

El Ahumado/Casas Viejas/Garita Chapina Region. This region, known as the Pasaco coast, supplied the modern department of Jutiapa, and possibly parts of Chiquimula.

As this reconstruction indicates, the trade of Pacific coast salt was highly localized and did not involve distances of more than 120 km.

These patterns have changed somewhat, since the modern salt works of San José and Champerico now supply the bulk of the salt to areas in the interior which received little or no coastal salt before the twentieth century, namely, the departments of Huehuetenango, Quiché, Izabal, Chiquimula, Baja Verapaz, Alta Verapaz, and Petén. These last two departments did not begin receiving coastal salt until the 1920s and 1930s; prior to that time they acquired English salt through Belize and the Golfo Dulce and salt from the Salinas de los Nueve Cerros, a salina in northern Alta Verapaz.

One fact that is apparent from the data is that, prior to the late nineteenth century, the coastal sources could not have met the demand of the highlands. If the salinas were producing 5,000 tons a year, as suggested, that amount would not have been sufficient to meet the demand for salt. It is for this reason, then, that the highland sources were so important.

THE HIGHLANDS

Amatitlán, Guatemala

The small salt-making industry of Lake Amatitlán is an affair of the past, a vague memory in the minds of a few elderly local residents. That such a memory still survives is surprising, since the salinas were abandoned almost a century ago.

Lake Amatitlán is a volcanic lake in a still volcanically active region. A number of underground springs (some boiling hot) feed into the lake, carrying with them a rich variety of minerals, including small quantities of salt (Deevey 1957). This salt accumulates on the beach at two locations on the south shore of the lake, at El Salitre and Playa Lavaderos (fig. 4.4).

This salt was once exploited by means of a sal cocida process, similar to that used on the coast, the only difference being the use of *coladeras* or *cajones* (raised boxes or hoppers) instead of canoes to leach the soil. As in earlier times on the coast, the brine was cooked in ceramic ollas. Only two firsthand accounts of this industry—both dating to the seventeenth century—have been published (Gage 1928:177; Fuentes y Guzmán 1932–33, I:256–57).

History. The Lake Amatitlán region is rich in archaeological remains with five sites reported along the south shore of the lake and large deposits of prehispanic remains located at a dozen underwater localities (fig. 4.4). The remains from these sites clearly indicate that the shores of the lake have been inhabited from Late Formative times to the present (Shook 1952; Borhegyi 1958a, 1958b, 1960, 1965; Mata Amado 1975; Brown 1977).

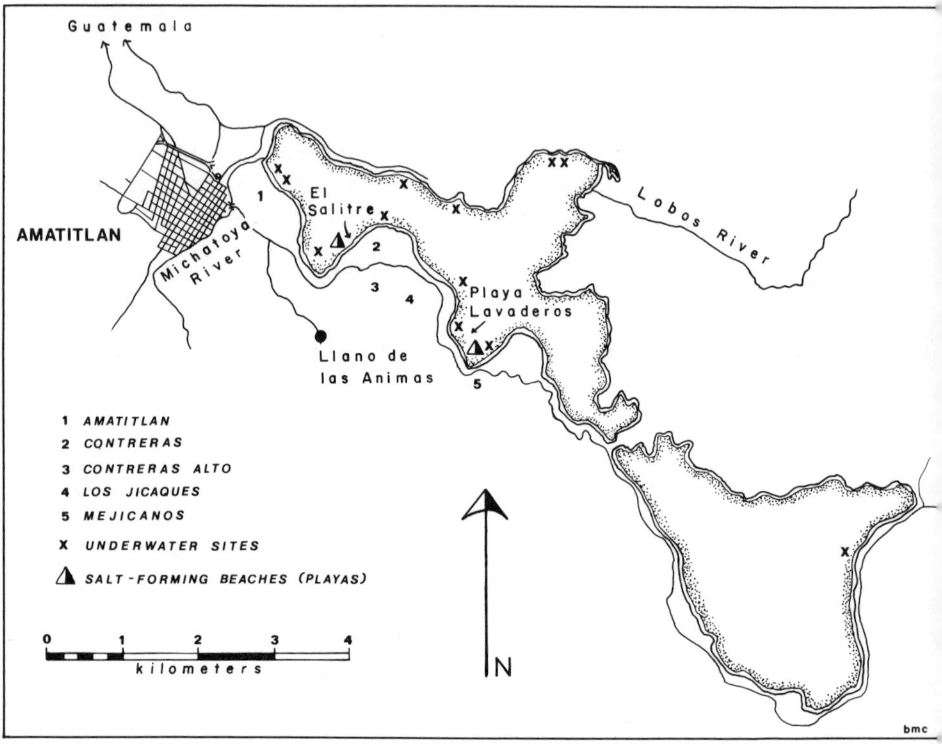

Fig. 4.4. Lake Amatitlán, Guatemala, showing location of salt playas and archaeological sites

Unsurprisingly, four of the archaeological sites are located next to the salt-producing beaches. While no concrete evidence of prehispanic saltmaking is available, the existence of such an industry is most likely (Feldman 1971:24, Brown 1977:277), probably with beginnings in Late Formative times.

The existence of saltworks at Amatitlán at the time of conquest is hinted at in the 1548 Tax List (AGI Guatemala 128), in which Amatitlán is reported to have paid a small tribute of salt.

Later reports of the salt works include the seventeenth-century accounts, as well as an early nineteenth-century report which notes that the natives of Amatitlán carried on a "considerable" trade in salt. The salinas were abandoned in the 1880s, when the railroad to the coast was built, and large quantities of cheap coastal salt began flooding the highland markets (Chinchilla Aguilar 1961:59–61).

In 1977, only a few old men from the village of Llano de las Animas remembered the saltmaking, which, along with fishing, once sustained their community. Borhegyi (1960:1) has suggested that these people are the descendants of residents of the prehispanic communities of Contreras Alto and Los Jicaques.

Production and Trade. Little reason exists to believe that Amatitlán ever produced any significant quantities of salt. The older men of Llanos claim that very little salt was produced; my own estimates, based on field observations and the potential of the technology used, lead me to believe that less than 10 tons a year were produced.

Such an amount would not go far. Informants said it was traded only to nearby communities. Consequently, I concur with Feldman (1971:185) that the trade of this salt was restricted to the Valley of Guatemala.

Sabalpop, Chimaltenango

Located a few kilometers west of the town of Patzún was the small hamlet of Sabalpop, obliterated by the 1974 earthquake. Near the town was a small spring that the natives exploited by means of a sal cocida process. The quantities produced were tiny, a few hundred kilograms a year, and were consumed by nearby households. I was unable to locate any historical record of this source. I learned of it from Robert Wauchope (personal communication 1976). Informants in nearby towns have only a vague recollection of this salina.

San Miguel Acatán, Huehuetenango

San Miguel Acatán, a small village of Kanjobal speakers, lies on the northwest flanks of the Cuchumatanes Mountains at an altitude of 1,800 meters. Eleven kilometers northwest of town, on the north side of the inner gorge of the Santa Catarina River, are three salt springs known as Atzam or Salinas San Miguel (fig. 4.5).

These salinas were exploited until the 1960s by people from San Miguel, who cooked the brine from the springs in small ceramic ollas similar to those used for cooking brine in San Mateo Ixtatán (see next section on San Mateo). According to local informants, relatively small amounts of salt were produced: from various accounts I gathered that the annual output was 5 to 10 tons.

This salt was sold at the market of San Miguel and at a few nearby markets, notably at Jacaltenango and Nentón (fig. 4.5). A large part of it was purchased for use on cattle ranches.

Although I was unable to locate any pre-twentieth-century references to this source, local informants claim that the salinas have been exploited since ancient times. In fact, the only published reports of this source merely mention its existence (Recinos 1913:48, 199; Quiñonez 1930:1926). Recinos (1913:239) reports a prehispanic site near the salinas, which raises the possibility of prehispanic salt exploitation.

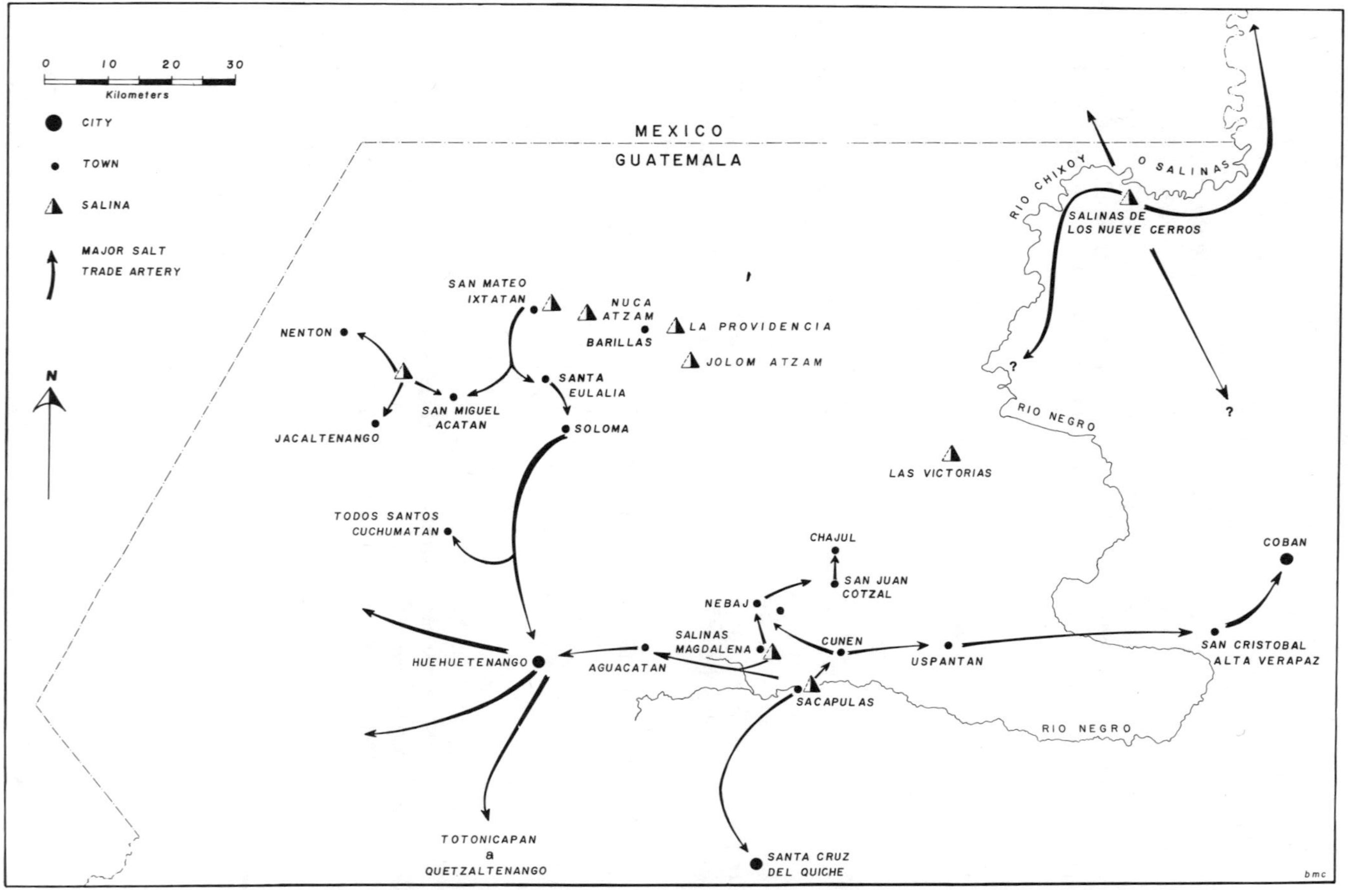

Fig. 4.5. Location of major towns and salinas, as well as past and present salt trade routes in northwest Guatemala

In the late 1950s a road was built into San Miguel and cheap coastal salt became available in the town market, putting an end to the local salt industry.

San Mateo Ixtatán, Huehuetenango

This is a well-known salt source, one of the two still being exploited in the Guatemalan highlands (the other is at Sacapulas). San Mateo Ixtatán is a small town of 5,000 Chuj speakers located in the northwest corner of the Cuchumatanes at 2,540 meters above sea level (fig. 4.5). Ixtatán means "the place of the salt," and, indeed there are several salt springs in the vicinity of San Mateo that have been tapped by wells, The main wells lie directly below the town on the south side of the Paijilá (or Los Laureles) River. Three of these wells are covered by small sheds; a fourth well, now abandoned, is surrounded by the foundation of a collapsed shed (fig. 4.6).

Fig. 4.6. View of the salinas at San Mateo Ixtatán, Huehuetenango, numbered to correspond with text

The salt-making process, which has remained virtually unchanged since prehistoric times, is extremely simple. Saltmaking is a job for the women of the town who take large *tinajas,* or water jars (figs. 4.7 and 4.8; see Reina and Hill 1978: 95, 109) down to the wells, which are open daily from 3 to 5 p.m. The wells are owned and operated by the municipality, and for 20 cents the women fill their jars with the saltiest brine I have ever tasted (the salinity was 39 percent). They then lug the jars up the steep hill—an arduous trek with a 25-kg load—to the town and then to their homes, where they make the salt by cooking the brine in small open vessels known as *apastes* (fig. 4.9). The cooking takes approximately twelve to fourteen hours; as the water evaporates, they continue adding brine, until a solid mass of salt is formed. As this salt cools and hardens, the apaste cracks into several fragments and is discarded. The final product is a solid 22-kg cake of salt known as a *pilón* (fig. 4.10). These are sold to wholesalers, who break them into smaller pieces for retail sale. Each of the approximately twenty to twenty-five households involved in saltmaking in 1977 was producing about two pilones a week.

Of the four wells, only two are now exploited. Well No. 1 is the main source of brine; it is open every day and has been in continual use since prehispanic times. Well No. 2, whose brine level is very low, is only opened every fifteen days. Local folk claim that it was once more productive. Well No. 3 was abandoned in 1977, as rising levels of fresh water had reduced the salinity of the brine to insignificant levels. I assume, however, that it was in use in the 1940s, for a report dating to 1940 alludes to the exploitation of these three wells (Stadleman 1940:95).

Well No. 4 has long been abandoned. Its shaft had collapsed, and only the foundations of the shed remained. Apparently this well was flooded with fresh water from the adjacent Paijilá River. I believe it was in use around the turn of the century, as two reports of that period mention the existence of four wells (Recinos 1913:213, Termer 1957:78–79).

The reduction of salinity is due probably to a rise in the level of the water table, a change which may have been caused by a rise in the river. The location of the wells would seem to support this explanation. Well No. 4 is the lowest and closest to the river, while Well No. 1 is the highest and farthest from the river.

There are several other salt springs in the vicinity of San Mateo. At the hamlet of Chixjoj or Chexjoj, 2 km northwest of San Mateo, are four salt springs, exploited in the early twentieth century. These have been reported in the literature as Santa Delia and Chilapchi (Recinos 1913:49, Quiñonez 1930:136). Recinos notes that the brine from these springs was cooked in peroles but that production was of "minor importance." Informants tell me that the salt from this source was used only by nearby cattle ranches owned by *ladinos* (nonnatives). These springs, which dried up many years ago, were finally buried in the 1976 earthquake.

Fig. 4.7. Women of San Mateo Ixtatán transporting brine in ollas to their homes to cook

Fig. 4.8. Tinajas, or water jars, used for transporting brine, set against the prehispanic wall that surrounds Well No. 1 at San Mateo Ixtatán

Fig. 4.9. An apaste, the vessel used for cooking brine, from San Mateo Ixtatán. (Penknife measures 8 cm.)

Fig. 4.10. A pilón of salt produced at San Mateo Ixtatán. (Lighter measures 8 cm.)

Another spring has been reported at Chilón, 2 km downstream from the San Mateo salt wells (Quiñonez 1930:136–37). This is a mere trickle of low salinity, and I found no evidence that it was ever exploited.

Quiñonez (1930:136–37) reports five other sources in the municipality of San Mateo: El Porvenir, Yola, Xul, Hul, and Santa Josefa. The first two are undoubtedly the same as Salinas el Porvenir, located in the Barillas region; Xul is a name given to one of the four wells at San Mateo (Stadleman 1940:95), and Hul is probably an alternate word for the same source; finally, I found no evidence of any source known as Santa Josefa.

Prehispanic History. San Mateo is an ancient town whose occupation dates far into the prehispanic past. A large archaeological site known as Calvario lies on the eastern edge of town, and the town itself sits on a vast midden of remains from various periods. A large portion of this midden is made up of sherds of apastes discarded from several centuries of saltmaking. The apastes used today are identical to those of prehispanic times, as I found prehispanic apaste sherds associated with Late Postclassic materials in several exposed stratigraphic cuts in and around the town.

Several prehispanic structures remain around the salt wells, including a well-preserved Late Postclassic temple platform located behind Well No. 1. The first two wells are surrounded by the remains of prehispanic walls (fig. 4.6). Moreover, I suspect that the foundations of all four sheds are also prehispanic, as are the well shafts and staircases leading down to the brine; some of the timbers lining the shaft of Well No. 1 are undoubtedly very old, for they are in the early stages of petrification.

A third complex of prehispanic structures, known as Catepan, lies on the northern flanks of the valley opposite the salt wells. Among these is another well-preserved Late Postclassic platform.

I have offered only a brief description of these remains, which have been the subject of several preliminary reports (Recinos 1913, 1954; Termer 1927, 1931; LaFarge and Byers 1931). Surface collections from all three sites and from in and around the town yielded Late Postclassic, colonial and modern sherds. The standing architecture of all three sites shows unmistakable Late Postclassic characteristics, very similar to those of the architecture of Zaculeu and other late highland sites.

There is also a hint of earlier remains: LaFarge and Byers (1931:22) note that Calvario has a number of architectural features in common with the Terminal Classic period sites of Tenam and Chinkultic in Chiapas. Future excavations will probably reveal evidence of Classic period, and perhaps earlier, occupations.

What is clear from the data is that San Mateo was definitely aligned with the cultural traditions of the western highlands during the Postclassic period. Whether or not San Mateo was part of the late Mam Kingdom, whose capital was at Zaculeu, is not clear (LaFarge and Byers 1931:6–7); the presence of massive Zaculeu-type architectural complexes in such a remote locality leads me to believe that such was the case. At any rate, the Mam must have depended on San Mateo for salt.

Nowhere in Mesoamerica have I ever felt so close to the prehispanic past as in San Mateo. The people, who speak one of the oldest known Maya languages, retain native dress and customs and make a living much as they did four centuries ago: growing crops, making salt, and manufacturing textiles (the ornate *huipiles* of San Mateo are famous throughout Guatemala). They have made few concessions to the modern world.

The most marked feature of the ongoing past in San Mateo is the aura of sacredness and ritual that pervades the entire town. There are more than twenty large wooden crosses in and near the town: on hilltops, on and around the prehispanic structures, and next to the salinas. As in Chiapas, the crosses often occur in groups of three; each of the first two salt wells, for example, has a cluster of three crosses in front of the shed doors. At the foot of these and other crosses one often finds small hearths where incense is burned.

A visitor to San Mateo cannot avoid being impressed by the incessant chanting and burning of incense that goes on from dawn to dusk. These rituals are carried out by older women who continually make the rounds of the crosses; they also make offerings at the ancient Catholic church in town, as well as on the tops of several of the prehispanic temple mounds. Such activities are clearly a continuation of prehispanic tradition, cloaked in a thin veneer of Spanish Catholicism (Vogt 1969:390; for lengthier discussions of ritual in the Cuchumatanes see LaFarge and Byers 1931; Oakes 1951; Grollig 1962).

As noted previously, several of the clusters of crosses are shrines dedicated to the ancestral deities who "own" and look after the salt wells, and the attendant rituals are designed to ensure that the wells remain productive; after all, the "sacred" salt has always been a major source of income for San Mateo. Such a combination of religion and economics is not Catholic but very much prehispanic.

Colonial History. It is not known when the Chuj of San Mateo first came into contact with the Spanish, though it was probably not long after Pedro de Alvarado captured Zaculeu in 1525. After the siege, he left a garrison at Huehuetenango with orders to subject the Mam country. By 1558 "San Mateo Ystatan" appears in Spanish documents, as being under the jurisdiction of the parish of Soloma (Recinos 1954: 6–7; LaFarge and Byers 1931:7).

The first colonial record of the salt industry appears in the 1548 Tribute List in which San Mateo is reported to have paid a salt tribute of 12 fanegas (1.4 tons); this was an extremely steep tax, comparable only to that of major coastal sources. Another sixteenth-century reference to the industry is that of Estrada and Niebla (1955:79) who briefly mention it in a 1579 document.

The first detailed description of the salinas appeared in 1690 (Fuentes y Guzmán 1932–33, III:98), and subsequent accounts were written in 1740 (Olaverreta 1935:19), 1763 (Pardo 1937:308), 1770 (Cortés y Larraz 1958:78–79; this reference mistakenly locates the salinas in "Zalama," i.e., Soloma), and 1798 (Hidalgo 1952:410–11). These reports provide a wealth of detail on production and trade.

Modern History. The available nineteenth-century reports on San Mateo do little more than report the existence of the salinas (Suasnavar et al. 1964:118; Juarros 1823:81, 486; 1936:48; Sapper 1897:304).

Twentieth-century accounts are far more informative, as they offer a variety of detail on the archaeology, history, and ethnography of San Mateo. The best accounts are those of Recinos (1913, 1954), LaFarge and Byers (1931:6-7; 216–26), and Termer (1927:18–19; 1931; 1941:225–26; 1957:78–79). Briefer reports include those of Stadleman (1940:95), Goubaud Carrera et al. (1944:16–21) and Petersen (1976:243).

Because of its isolation, San Mateo was largely unaffected by events of the nineteenth and twentieth centuries. In the early 1960s, a dirt road from Soloma and Huehuetenango reached San Mateo, bringing about perceptible change in the town. The influx of packaged coastal salt has brought about a decline in the local salt industry, which was already being affected by dropping brine levels. The lost salt revenues are now being replaced by income from the sale of timber and textiles, whose export is facilitated by the new road.

Production. In the late 1970s, San Mateo was producing approximately 50 tons of salt a year. Production was once higher. Town officials estimate that during the 1930s and 1940s, when three of the four wells were producing brine, the output exceeded 80 tons a year. At the beginning of the century, when all four wells were being exploited, annual production was approximately 92 tons (Recinos 1913:48). An identical figure has been reported for 1798 (Hidalgo 1952:410–11), which leads me to believe that such a sum is representative of the output of the industry prior to modern times.

As for the nearby springs at Chixjoj, all informants concurred that production there could not have exceeded 10 tons per year.

In sum, the combined production potential of the San Mateo area appears to have been around 102 tons a year, or sufficient salt to supply the basic needs of approximately 35,000 people.

Trade. While the present trade of San Mateo salt is limited to nearby communities in the northern Cuchumatanes Mountains, its past trade range was far more extensive.

Until the 1960s, the traditional range was the entire department of Huehuetenango. Informants throughout this region confirmed that on Sunday and fiesta day markets, San Mateo salt sold in Barillas, Quetzal, Santa Eulalia, San Juan Ixcoy, Soloma, Nentón, San Rafael Petzal, San Miguel Acatán, Todos Santos Cuchumatán, Chiantla, and Huehuetenango. It was alleged to be widely sought because of its purported medicinal properties (an analysis of the brine revealed high levels of iron, calcium, and magnesium, which may explain this reputation).

During the Colonial period, the trade range was far more extensive. Several accounts of the seventeenth and eighteenth centuries note that San Mateo salt was traded on a normal basis to Totonicapán, Quetzaltenango, and the "llanos de Chiapas" (just across the present-day border) (Fuentes y Guzmán 1932 – 33, III:98; Olaverreta 1935:19; Pardo 1937:304; Hidalgo 1952:410 – 11).

This pattern of marketing San Mateo salt as far south as Quetzaltenango appears to have continued into the early nineteenth century (Juarros 1823:81), although it eventually came to a halt as coastal salt began to spread into the highlands.

By the 1940s, coastal salt was being traded into the Cuchumatanes, and the San Mateo industry began to feel the effects of the competition (McBryde 1947:73 – 74).

In the late 1970s I found San Mateo salt for sale in only a few nearby markets: Barillas, Santa Eulalia, Soloma, and San Miguel Acatán. I was also told that small amounts of it were still sold on special fiesta days in Chiantla and Huehuetenango, where its medicinal properties and unusual taste attracted a few customers. It was also bought for livestock, for whom it was considered to be highly beneficial. It generally sold for ten cents to fifteen cents a pound, while coastal salt sold for five cents a pound. At such a price, one can only wonder how much longer the San Mateo industry will survive.

Barillas, Huehuetenango

In the vicinity of the town of Barillas are three salt springs which were exploited in the nineteenth century and subsequently abandoned. Little information is available on these sources, whose combined production was less than 50 tons a year. As far as I have been able to ascertain, this salt did not circulate in the native market economy, but was produced for the sole consumption of livestock at ladino cattle ranches in the Barillas region. I was also unable to determine if any of these sources was exploited in prehispanic or colonial times. The following data on these sources were gathered mostly from informants.

Nucá Atzam. Also known as El Quetzal, El Progreso, Yulatizu, or Salinas El Porvenir, this spring was located 1 km west of the town of El Quetzal, in the vicinity of the hamlets of Nucá and El Progreso and of the Yulatizú River. This spring was tapped in the 1880s by a ladino operator who cooked the brine in iron vats. At this time the operation was known as Salinas El Porvenir.

The source was also exploited in the early years of the twentieth century (Recinos 1913:215, Quiñonez 1930:136, Stadleman 1940:98), but was abandoned in the 1940s. According to local informants, an erosion-induced landslide mixed the spring brine with fresh water, reducing its salinity to insignificant levels. Litke (1975:84–85) has suggested that the salt deposits here were either leached out or mined out. I doubt they were mined out, for we have no record of salt mining anywhere in the Maya area.

La Providencia. This spring, located 6 km east of Barillas near the ranch of La Soledad (also known as Los Nogales), was also tapped by a rancher in the late nineteenth century. As in the case of Nucá Atzam, this spring, whose brine was cooked in iron vats, was obliterated by a landslide in the 1930s. One informant told me that large chunks of rock salt were found in the landslide debris, suggesting that the salt deposits may lie near the surface.

This source has been reported by Recinos (1913:49), Quiñonez (1930:136), and Litke (1975:84–85).

Jolom Atzam (Yulconob). Located near the village of Amelco, 9 km southeast of Barillas, this spring was reportedly exploited during the nineteenth century by area natives who took the brine home and cooked it in ceramic ollas. The salt produced was strictly for local household consumption.

In the late nineteenth century this source became part of a ladino ranch, and several apparently unsuccessful attempts were made to start a sal cocida operation. The source has been reported by Quiñonez (1930:136).

Other Reports. Quiñonez (1930:136) also reports a salt source at Cerro Chiblac, located 11 km east of Barillas, but I was unable to confirm its existence. As I visited Barillas only briefly, my data on this area are scant. I made only a cursory inspection of Nucá Atzam and failed to visit the other sources, because the road east of Barillas was washed out. Time limitations prevented me from returning to Barillas. There is clearly a need for further reconnaissance.

Sacapulas, Quiché

Sacapulas is a Quiché town of approximately 14,000 people, located at the foot of the Cuchumatanes Mountains 33 km north of Santa Cruz del Quiché (fig. 4.5). It lies on the southern flank of the Río Negro Valley at an altitude of 400 m.

The salinas of Sacapulas, continuously exploited from prehispanic times to the present, are famous throughout Guatemala for the black salt they yield, which is believed to possess various medicinal properties. The salinas are located below the town along the south shores of the Río Negro. The salt-making technology, basically a sal cocida process similar to that of the Pacific coast, has not changed in centuries and is the most complex and laborious I encountered in the survey.

The salinas are a series of salt playas, flat, salt-encrusted earth, exposed by erosion of alluvial soil and rocks. Next to the river, these flats are almost flush with the surface of the ground; farther away, as much as 2 m of alluvial deposits must be cleared to reach the playa soils.

This saltmaking begins in the morning with the scattering of a layer of fine black soil over the salt playas. At intervals until mid-afternoon, hot water from nearby springs is sprinkled over the soil. At the end of the day the soil is raked into small piles. The same process is repeated for three days, during which time the soil, which has been repeatedly moistened with water and dried by the sun, has absorbed a fair amount of salt from the playa soils. Since intensive solar heat is essential, this process may take longer if cloudy days intervene. This factor also limits the leaching of the playas to the dry season.

On the fourth day, the soil is carried in baskets up the river bank to the cocinas or cooking houses (there are fourteen), where it is either stored or immediately processed. Processing involves placing salt-impregnated soil in cajones or coladeras, which serve the same purpose as the canoes on the coast (fig. 4.11). The cajones are slat-bottomed boxes upon which are laid a petate, or reed mat, which serves as a filter. The soil is then soaked with slightly saline water from nearby brine wells, which collects in a tile pan below the cajon.

In the cocinas the brine is cooked over a fire in ceramic ollas similar to those used at San Mateo. Usually the salt is cooked until it hardens; the resulting half-hemisphere is then joined to one from another vessel to form a *bola,* or ball of salt, that weighs from 10 to 20 kg. The brine used for making the bolas contains a fair amount of playa soil, and minerals in the soil are partly responsible for the salt's dark color.

Fig. 4.11. A cajón used for leaching playa soils at Sacapulas, Quiché. The brine is collected in a tile pan beneath the cajón.

In some cases, the salt is removed from the vessels before it hardens and is placed in wooden molds in the shape of flat, circular cakes, which are then dried in the sun. These cakes, known as *quesitos,* weigh approximately 6 ounces.

An interesting feature of the salt-making process of Sacapulas is the use of cylindrical river cobblestones in the floor-level fireplaces in the cocinas. These stone cylinders are used to prop up the brine-cooking vessels over the fire. Such cylinders are not unique to Sacapulas. Similar clay cylinders have been reported from Late Formative salt-making sites along the Pacific coast, at Ocós (Coe and Flannery 1967:91, pl. 28) and Salinas de Acapán (Shook personal communication 1977). I also located a clay cylinder on the river bank at Sacapulas that was associated with Late Postclassic ceramics.

Several authors have described the saltmaking of Sacapulas, the most informative accounts being those of Dollfus and Mont-Serrat (1868:228 – 31), Burbank (1939:192), McBryde (1947:58), Judson and Judson (1949:122), Helfritz (1963), Pettersen (1976), and Reina and Monaghan (1981). The last three are the most comprehensive.

Prehispanic History. The Sacapulas region is rich in archaeological remains. Overlooking the salinas from a bluff on the north side of the Río Negro is the prehispanic site of Chutinamit, occupied during Early and Late Postclassic times (Smith 1955: 20–21; Fox 1978:77–81).

Sacapulas itself is also of considerable antiquity, and prehispanic midden materials are found scattered about the town. It was known as Tuhala (also Tuhal or Tajul) in prehispanic times. Abundant midden remains are also found around the salinas, particularly in the river embankment below the cocinas. A collection of sherds from the embankment yielded Late Classic and Late Postclassic materials.

Sacapulas lies in a strategic position at the crossroads of major trading routes. Its location along the Río Negro (fig. 4.5) not only placed it in an advantageous position for regional east-west trade, but also gave it access to the Atlantic lowlands, since the Río Negro is the main tributary of the Usumacinta drainage. On a north-south axis, Sacapulas lies between the eastern Cuchumatanes and the central Quiché region, and was a logical node for interaction between these areas.

Sacapulas not only possessed a critical resource but also acted as a commercial center of considerable importance because of its position at the crossroads of some of the most important trade routes of the central highlands. In fact, it has been suggested that Sacapulas was a major regional capital in prehispanic times (Fox 1978:111). Its importance as a prominent trading center continued well into the Colonial period (Ximénez 1929–31, I:191 [Bk II, Ch. 13]).

Undoubtedly, a desire for control of the salinas as well as the trade routes led to the conquest and incorporation of Sacapulas into the Quiché state in the fourteenth century (Fox 1978:111). It was still under Quiché control when the Spanish arrived in the sixteenth century.

Colonial and Modern History. There is considerable documentation on Sacapulas and its salt industry from the conquest to the present time. An important document on the pre-conquest history and mythology of Sacapulas is the *Título de los Señores de Sacapulas* (AGCA 1551), in which continual reference is made to the salinas. This document makes it clear that the salt works were a focal point of the legendary, political, and economic history of Sacapulas long before the conquest. Although several transcriptions and preliminary studies are available (Crespo Morales 1968; Acuña 1968; Feldman 1972; Carmack 1973:37–39, 355–59), a thorough analysis of this document has yet to be made.

The earliest detailed account of Sacapulas saltmaking was written in 1635 (Tovilla 1960:218–19), which indicates that the salt-making process has changed little since then.

Another valuable source of information on the Colonial period industry is a large collection of unpublished documents dating to the late eighteenth century, which concern litigation over ownership of the salinas. These documents contain important data on the social organization and landholding patterns of the time. One document delineates individual properties in the salt playas, and names ten of the property owners (AGCA 1779). Altogether, the documents indicate that control of the salt works has remained in the hands of the two main residential groups of Sacapulas from prehispanic times to the present (Hill personal communication 1981).

Although Sacapulas has not changed greatly over the centuries, the twentieth century has witnessed a pronounced decline in its salt industry. This decline began with the introduction of coastal salt into the region in the 1940s, a process that undermined many of the highland industries. Another event which seriously affected production was the Río Negro flood in September 1949, which deposited several thousand tons of rocks, boulders, and alluvial sands over the salt playas (Pettersen 1976:55). Informants told me that the flood buried more than 60 percent of the salt playas.

Production. Since saltmaking at Sacapulas is a cottage industry without records, it was difficult to arrive at estimates of production. I interviewed several saltmakers, saltpeddlers, a few village elders and the alcalde, whose estimates ranged from 5 to 12 tons a year. Another recent estimate places production at 9 tons a year (Lefond 1969:138). For convenience, I will use the figure of 10 tons a year.

Everyone I talked to noted that prior to the flood and the arrival of coastal salt, production was four to five times what it is presently, or 40 to 50 tons a year. In 1940 McBryde (1947:73) observed seventy-five salt vendors at the weekly Sacapulas market, each selling between 25 and 50 pounds. He must have been there during the salt-making season, since such a quantity of salt would not have been available during the off season. Taking this factor into account, as well as certain other modifying factors, I have made several different calculations based on his figures: all the estimates fall between 37 and 66 tons a year.

In sum, I believe that the maximum potential production of Sacapulas in the past did not exceed 70 tons, which would have supplied the basic needs of approximately 24,000 people.

Trade. The modern trade range of the salt from Sacapulas is limited; after all, 10 or 12 tons will not go far. Most of the salt is consumed at nearby cattle ranches, and some is sold in the market place of Sacapulas, to local and regional customers, and to visiting outsiders who have an interest in the exotic "black salt of Sacapulas."

Beyond this local marketing, some of the salt is carried to other regional markets on special fiesta days. These include Aquacatán and Huehuetenango to the west, Santa Cruz del Quiché to the south, Cunen and Uspantán to the east, and Nebaj, Cotzal, and Chajul to the north (fig. 4.5).

This "fiesta trade range" is a skeletal vestige of past trade. According to informants at Sacapulas, these markets were supplied with Sacapulas salt regularly before 1940. Most of the salt went south into central Quiché, a pattern which undoubtedly dates to at least the fourteenth century when Sacapulas was integrated into the Quiché Kingdom.

The central Quiché area probably became heavily dependent on Sacapulas salt after the Cakchiquel revolt in the fifteenth century. The Cakchiquel, who live between central Quiché and the coast, may have cut off or reduced incoming salt from coastal sources.

The trade of Sacapulas salt to the west is apparently also very old: Huehuetenango paid a salt tribute—of Sacapulas salt—in 1548 (AGI Guatemala 128). The trade to the east also dates to the sixteenth century: a report of 1574 notes that the province of Verapaz, whose capital was at Cobán, depended almost entirely on salt from Sacapulas (Viana et al. 1955:21–22). The road to Cobán was still a major trade route in the nineteenth century (Stoll 1886:407), as it is today.

The northern trade into Ixil country (toward Nebaj) is also ancient, most likely dating to prehispanic times. Given its chronic shortage of agricultural lands, Sacapulas probably relied on this trade to acquire additional maize; such a pattern was "old" in the eighteenth century (Hill personal communication 1981).

The Ixil people of northern Quiché had access to two other salt sources, Salinas Magdalena and Las Victorias, which may have supplied some of their needs; they may also have acquired some salt from the Barillas area to the northwest (see fig. 4.5 and sections below). I find it unlikely, however, that these small sources could have met the demand of the entire Ixil country, which, ultimately, would have had to rely on Sacapulas for a fair amount of its salt.

Salinas de Magdalena, Quiché

This town, known during the Colonial period as Santa María de Magdalena, is located 6 km northwest of Sacapulas in the eastern fringes of the Cuchumatanes Mountains (fig. 4.5).

The salinas, which consist of five tiny brine springs, are located approximately 1 km southeast of town at the bottom of the gorge of the Río Pajaritos, just below a prehispanic site. These springs, mere trickles in 1977, were formed when a landslide altered a then single spring that was the locus of a large sal cocida operation in the late nineteenth century.

Prehispanic History. The inhabitants of the prehispanic site, which literally sits above the salt springs, were undoubtedly involved in saltmaking. First reported by Stoll (1886:385–87), the springs include extensive midden remains dating to Late Classic, Late Postclassic, and Colonial times.

Colonial and Modern History. During the early Colonial period, the community became Santa María de Magdalena. Although I have not located any published information on it, a set of eighteenth-century documents sheds some light on the village's fate during the Colonial period. Apparently devastated by an epidemic during the 1630s, the town was subsequently abandoned. Survivors moved to Sacapulas and Cunén. One hundred and fifty years later the lands around Magdalena became the subject of litigation between the governor of Sacapulas and the survivors' dependents. It is not clear from the documents how the matter was resolved (AGCA 1786–90).

Since I found no other reference to the town before the late nineteenth century, I believe that it remained abandoned. Moreover, it appears that Ixil natives from nearby Nebaj laid claim to the salinas, where they were collecting brine in the 1880s (Stoll 1886: 385–87).

In 1881 the land around Magdalena was acquired by Rufino Barrios, then president of Guatemala, who set up a large sal cocida operation at the spring. A large flume was built to tap the spring and conduct the brine to a flat area outside the gorge. A cocina was constructed where the brine was cooked in large hornos over adobe ovens. This operation has been described in detail by Manó (1940) and Stoll (1886:385–87; cf. also Sapper 1897:188, 304; Termer 1957:78).

Barrios's operation was abandoned in 1885 when he died. The project's failure was due to technical difficulties (the flume was repeatedly broken by landslides), as well as an eventual scarcity of firewood. The operators also encountered difficulties transporting the salt to market over the region's rugged terrain (Stoll 1886:385–87; Villacorta 1883).

Scarce firewood was probably the most important factor in its demise. The salinity of the spring was so low (approximately 4 percent) (Stoll 1886:385–87) that huge amounts of brine had to be cooked to acquire significant quantities of salt. The operation consumed 1.4 tons of wood a day (Manó 1940:510)! Lacking any notion of environmental impact, the operators had deforested a vast region and had irreversibly deprived themselves of fuel.

In 1977 all that remained of the salt works were the walls of the cocinas and several related structures. Two iron vats also survived as water troughs in the main square of the modern village of Magdalena. The people of the village remember little about the salinas, but one ugly reminder survives: the scars of the deforested countryside.

Production and Trade. According to two independent sources, the annual production of Barrios's operation was between 73 and 110 tons (Manó 1940:503; Stoll 1886:385). I have been unable to trace the trade of this salt, but I suspect it was sent to Huehuetenango, since a road was constructed in the early 1880s between Magdalena and Río Blanco, in the direction of Huehuetenango (RG, MF, *M*).

It is difficult to estimate the production of the native industry prior to the 1880s, but it must have been considerably less than the output of the relatively modern Barrios operation. I have arrived at a production estimate of 12 tons a year based on comparisons with other similar sources in the highlands, as well as discussions with Sacapulas saltworkers familiar with Magdalena. This amount would have supplied approximately 4,000 people and was probably traded to nearby communities in northern Quiché.

Las Victorias, Quiché

Located approximately 30 km due north of San Miguel Uspantan is the cattle ranch of Las Victorias, which has a small brine spring. In the late nineteenth century this property was acquired by Germans, who reportedly cooked the brine in iron vats and supplied nearby cattle ranches with salt. The operation continued until World War II, when the lands were confiscated. The spring has since been abandoned.

I did not visit this source, but gather from informants in Uspantan that the amounts produced were very small, definitely less than 10 tons a year. I was unable to locate any other information on this source.

THE LOWLANDS: SALINAS DE LOS NUEVE CERROS

This legendary salt source, known as Bolontehuitz in Yucatec Maya, is located on the Chixoy River at the northern edge of the department of Alta Verapaz (fig. 4.12). Although some authors have placed this source in the highlands, Thompson (1970) and Dillon (1977) have shown, quite conclusively, that it is a lowland source.

The source consists of a large brine spring of high salinity which flows out of a gorge on the west side of Cerro Tortugas onto some flats at the base of a range of hills known as Nueve Cerros. From the flats, the brine drains into the Arroyo Salinas, which empties into the Chixoy River; at this juncture, the Chixoy becomes the Salinas River, a name it retains until it joins the Pasión and becomes the Usumacinta. For a detailed description of the Nueve Cerros region, the reader should consult Dillon (1977).

The brine spring of Nueve Cerros has been exploited from prehispanic times until the present century by means of sal cocida operations, the last of which ceased in 1937. Most of our information on Nueve Cerros is based on archaeological and historical data.

Prehispanic History

The remains of the large prehispanic settlement of Bolontehuitz have been the subject of intensive archaeological investigations directed

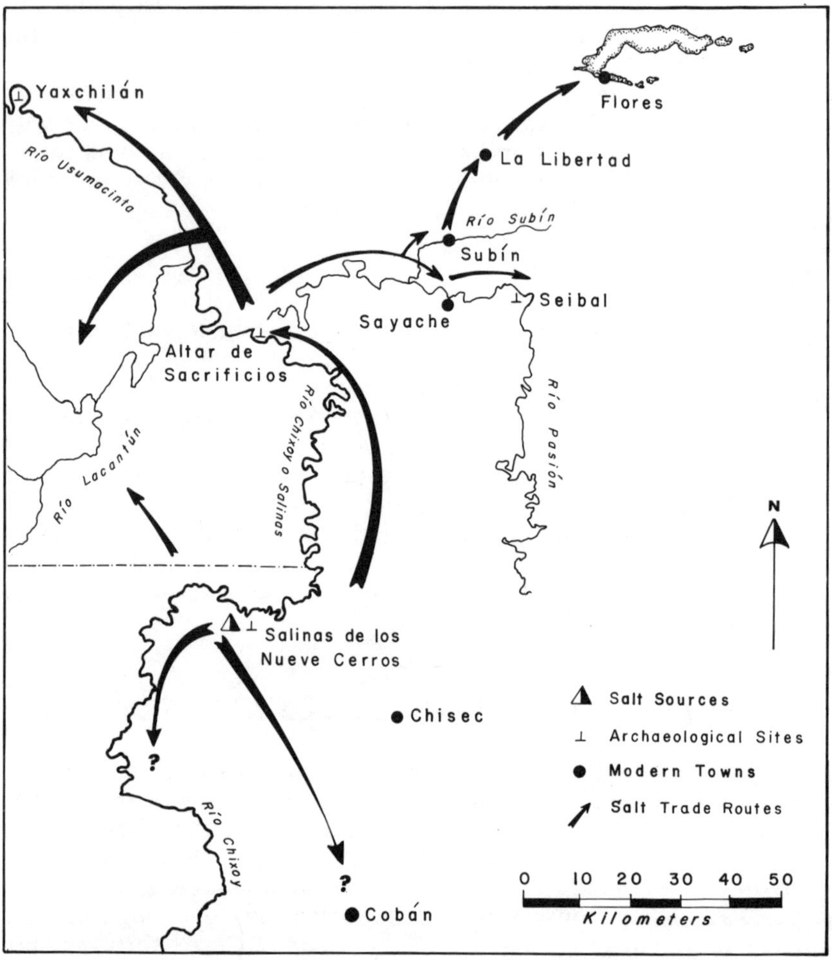

Fig. 4.12. Salinas de los Nueve Cerros region, showing major salt trade routes of the late nineteenth and twentieth centuries

by Brian Dillon between 1975 and 1978 (Dillon 1977, 1979). Since his reports contain all the available data on the prehistory of the site, I will only summarize them.

The site of Nueve Cerros was first reported by Habel (1878:8), who visited the area in 1863. Subsequent reports deal with sculpture and artifacts from the site, but no other firsthand report of the ruins appeared prior to Dillon's arrival in 1975.

Dillon's excavations have revealed that Nueve Cerros was occupied from Late Formative through Terminal Classic times; he found no remains that could be assigned to the Postclassic period.

Evidence of saltmaking—vessels for cooking and storing brine (fig. 4.13) —dates from Early Classic times on. While definite evidence of earlier saltmaking is not available, the presence of a Late Formative settlement next to the salt spring strongly suggests that such activities were underway then.

Fig. 4.13. Classic period brine storage vessels at Salinas de los Nueve Cerros, Alta Verapaz, Guatemala. (Photograph courtesy of Brian Dillon)

An interesting feature of the archaeological record at Nueve Cerros is the distinct lowland character of the ceramics, architecture, and sculpture. It is clear from these remains that Nueve Cerros had strong ties with the Classic period populations of the central Petén; thus, Nueve Cerros was not only geographically in the lowlands, but culturally as well (Thompson 1970:126; Dillon 1977, 1979).

Despite the lack of evidence of a Postclassic occupation at Nueve Cerros, it is probable that the salinas were periodically exploited by salt-making expeditions from the neighboring Petén and Lacandón regions, as was the case during the Colonial period. In fact, Bolontehuitz was a known locality in Postclassic times, for it is reported in the Chilam Balam of Chumayel to have been one of the points of departure of the legendary Itzá migration of Terminal Classic/Early Postclassic times (Roys 1933:121; Thompson 1970:23).

Colonial History

Little is known of the colonial history of Nueve Cerros, for the region was beyond the sphere of Spanish control during most of that period.

As noted, Nueve Cerros was visited by native expeditions which went there expressly to acquire salt. One such expedition, of 140 Lacandón Indians was, in fact, observed there around 1630 (Tovilla 1960:211; Thompson 1970:135; Rathje 1972:391). This pattern of sporadic visits apparently continued until the nineteenth century (Habel 1878:5).

It appears that the salinas were of interest not only to the Lacandón, but also to the Itzá of the central Petén. According to an early colonial account, the desire to control the salinas often led to wars between these two groups (Dieseldorff 1909:866).

The Spanish discovered the salinas in 1625, and the report of their first expedition included a proposal to build a road between Yucatán and Guatemala via Nueve Cerros, which would have provided a means of transporting the salt to regional markets (Pinelo in Tovilla 1960:263–71).

Although government documents refer to an actual road to Nueve Cerros, which underwent repairs in 1880 (RG, MF, *M*1881), a road as such was never built. A trail, however, was eventually cleared between Cobán and Nueve Cerros. Since a late nineteenth-century map of Nueve Cerros designates this trail as a "camino real," I suspect it was set up during the Colonial period (Municipio de Coban 1882–1911). Habel, who traversed the trail in 1863, noted that it had bridges and rest stations (1878:5).

While I was unable to document Spanish exploitation of Nueve Cerros, the existence of a "camino real" leads me to believe that it was likely; after all, the Spanish were aware of the profit potential of the salinas (Ximénez 1929–31, 2:215 [Bk II, Ch. 69]).

Modern History

During the nineteenth century, Nueve Cerros became the property of the municipality of Cobán, and the salinas were exploited on a concession basis by a series of ladino operators. These operations were underway by the 1860s (Habel 1878:5–6; Sarg 1939:12), if not earlier, and continued until 1937 (Municipio de Cobán 1882–1911; Sapper 1897:74; 1901:176; Flavio Ponce personal communication 1978).

According to several accounts, the brine from the spring was cooked in iron vats (Villacorta 1883; Sapper 1897:188), whereas in earlier times it had been cooked in ceramic vessels. Col. Flavio Ponce, the last and only surviving operator of the salinas, told me that a previous concessionnaire had attempted on two occasions to produce solar-evaporated salt in the flats, but had been rained out both times.

While Ponce (personal communication 1978) felt that solar evaporation was not possible at Nueve Cerros, Dillon has proposed the possibility of a partial use of solar evaporation to increase the salinity of the brine before it was cooked (Dillon personal communication 1978). This would explain the function of the large numbers of huge brine storage vessels found at the prehispanic site. Kept under cover, the brine in these vessels could have been saved for rainier periods. A report of 1625 which refers to the salt works as "flats" (Pinelo in Tovilla 1960:263) leads me to believe that partial solar evaporation may also have been employed during the Colonial period. Such a procedure would explain the reported "employment" in 1630 of 140 people to maintain the salt pans and transport the brine.

As was the case with many highland salinas, the salt industry of Nueve Cerros came to an end with the advent of coastal salt. Col. Ponce had sold the Nueve Cerros salt downriver, mostly in the central Petén area, as did his predecessors. In the 1930s coastal salt began to reach this area, via Cobán-Sebol-Sayaxché-Flores, and in 1937, no longer able to compete, he closed his operation (Ponce personal communication 1978).

Production

When Habel visited Nueve Cerros in 1863, he was told that the daily output was 2,500 pounds (Habel 1878:5). I suspect that this figure is vastly exaggerated, as later production reports are much lower. Ponce also questioned this figure. Another report of 1883 notes that production was less than 10 quintales a day, or 168 tons a year (RG, SF). Karl Sapper, who visited the salinas in 1891, noted that production was approximately 100 tons a year (Sapper 1901:176).

The later figures are well in line with those of Ponce (personal communication 1978), who claims to have produced more salt at Nueve Cerros than any previous concessionnaire. In a good year his production reached 184 tons. Given the best possible circumstances, he estimates that the maximum that could have been produced at Nueve Cerros was 276 tons a year. Thus, I believe it safe to assume that in the nineteenth and twentieth centuries the production of Nueve Cerros never exceeded 300 tons.

Production in prehispanic and colonial times, when salt was cooked in ceramic vessels, must have been considerably lower. Although I have no way of estimating this output, I feel reasonably secure that it never surpassed the modern production of 300 tons a year.

Trade

On the basis of available archaeological and historical evidence, it would appear that throughout history the bulk of the salt produced at Nueve Cerros was traded north, into the lowlands of Chiapas and the central Petén (fig. 4.12). That this was the case in prehispanic times is clearly suggested by the strong cultural affinities between Nueve Cerros and sites in the Usumacinta and Pasión drainages (Dillon 1977, 1979). This trade continued into the early Colonial period, when the source was exploited by the Lacandón of Chiapas and the Itzá of the central Petén.

Such a pattern of trade is not at all surprising, since neither the Lacandón forest of Chiapas nor the Petén has any sources of salt. Moreover, the salt of Nueve Cerros was easily transported into those areas along the waterways of the Usumacinta River drainage.

This trade continued into the late nineteenth and early twentieth centuries, because most operators of the salinas found a ready market for the salt downriver. Col. Ponce provided a detailed outline of his trading activities, which were essentially like those of previous concessionnaires. He sold some salt locally to people who came to Nueve Cerros from hamlets and ranches in nearby Alta Verapaz, northern Quiché, and southern Chiapas. He transported the bulk of his salt down the Salinas and Usumacinta rivers, selling it to either the east or west. The eastern trade went via the Pasión River to Sayaxché and beyond, into the Petén. Some salt was taken up the Subín River to the town of Subín, and then transported overland to Flores. His western trade took him farther down the Usumacinta and up the Lacantún River, into the heart of the Lacandón forest.

Ponce never went farther north than the confluence of the Lacantún and Usumacinta rivers, but told me that other concessionnaires had occasionally taken salt as far north as Yaxchilán.

I believe that these trading patterns replicate the trade of prehispanic times. As an aside, it is worth noting that one prehispanic community held the key to this trade—namely Altar de Sacrificios, located at the confluence of the Usumacinta and Pasión rivers. We can only speculate whether Altar was a middleman, ally, or outpost of Nueve Cerros, but whatever its role, it was a crucial one.

Because of its position on a major navigable waterway at the southern edge of the lowlands, Nueve Cerros may well have been an important trade node between the highlands and lowlands. Although it had strong cultural affiliation with the lowlands, Dillon's investigations have revealed significant artifactual and architectural evidence which suggests that Nueve Cerros was also in close contact with the northern highlands of Guatemala (Dillon personal communication 1978).

An important item of any trade between Nueve Cerros and the highlands would have been salt, but I have found no conclusive evidence that Nueve Cerros salt was traded into the highlands in either prehispanic or early colonial times. Thompson (1970:135, 149–50) and Feldman (1978:14–15) have speculated that such a trade existed, but they lack solid documentation. Moreover, we know that shortly after the conquest the highlands of Verapaz depended solely on Sacapulas for their salt supply (Viana et al. 1955:21).

During the later Colonial period, some salt from Nueve Cerros may have reached Cobán, especially if the "camino real" between the two places already existed. This road was definitely in use in the late nineteenth century, at which time limited quantities of Nueve Cerros salt were being traded south into Alta Verapaz (RG, MF, *M* 1881; Sapper 1901:176).

Informants at Cobán told me that before 1920 occasional salt shipments from Nueve Cerros were brought in by mule trains. Nonetheless, during the nineteenth and early twentieth centuries, the Verapaz depended almost entirely on imported British salt from Lake Izabal on the Atlantic coast. Nor was Nueve Cerros able to satisfy the demand in the Petén, which imported sizeable quantities of British salt from Belize until the 1930s (Villacorta 1883; Sapper 1901; RG, SF 1884, II:9; RG, MF, *M* 1921; RG, MHYCP, *M* 1934). The scope of the British salt trade was surprisingly extensive. During the nineteenth century it was traded to Belize, Izabal, Alta Verapaz, Petén, and southern Quintana Roo. It appears, then, to have filled the vacuum created by the early colonial breakdown of the Yucatecan east coast salt trade. The British salt trade came to an end when salt from the Pacific coast of Guatemala began to reach the lowlands in the 1920s and 1930s.

5.

The Salinas of El Salvador

EL SALVADOR HAS A GROWING SALT INDUSTRY, the most modern and productive of Central America. Since the country lacks inland sources, this is and always has been a coastal industry (fig. 5.1).

Saltmaking in El Salvador has evolved in the same manner as did the industry of the Guatemalan coast. Prior to the mid-nineteenth century, all the salt works consisted of sal cocida operations in which soil was leached in canoes and the resulting brine was cooked in ceramic vessels. This process had been in use since prehispanic times.

The late nineteenth and early twentieth centuries saw a number of innovations. The ceramic vessels were replaced by peroles, which in turn were replaced by iron vat hornos. Asoleaderos, or solar evaporation tanks, were also built, thus eliminating the soil-leaching process. In 1978 in eastern El Salvador many sal cocida operations were still using asoleaderos and cooking the brine in iron vat hornos.

The sal cocida operations, however, are no longer the main producers. After solar evaporation technology was introduced in 1946, dozens of complexes of solar pans with tile floors appeared in the eastern part of the country. These operations now account for the bulk of national salt production.

Because of the competition from the modern industry, the sal cocida operations are gradually disappearing. There are no longer any salinas in the western part of the country. Most were abandoned in the 1950s and 1960s. The remaining operations in the eastern regions either have gone out of business or are converting to solar technology.

El Salvador, now the major salt-producing nation of Central America, exports large quantities to neighboring countries. Before the twentieth century, however, production was lower and the salt trade was strictly internal, with coastal salinas supplying only neighboring inland areas. The absence of evidence of a long-distance salt trade during the Colonial period leads me to believe that El Salvador did not supply other regions of the Maya area with salt during prehispanic times. Nineteenth-century production records confirm that such a trade would have been impossible: there simply was not enough salt to go around.

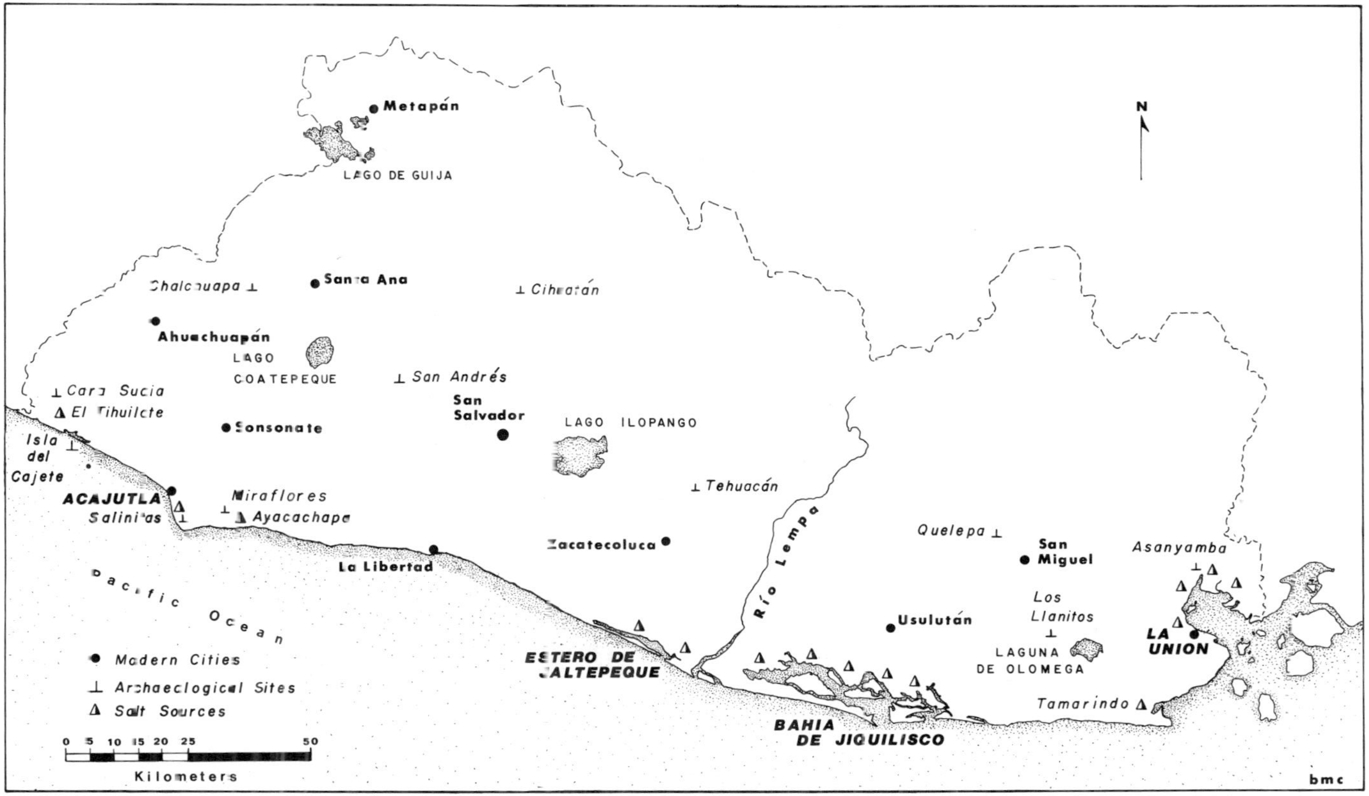

Fig. 5.1. El Salvador

A detailed account of the salt industry of El Salvador, which includes archaeological, historic, ethnographic, and modern industrial data, will appear elsewhere (Andrews n.d.). I will summarize that report here on a region-by-region basis.

ACAJUTLA REGION

Though no longer producing salt, the Acajutla area once had several salt-working localities at El Tihuilote (also known as Las Salinas), Salinitas, and Salinas de Ayacachapa (fig. 5.1).

All were sal cocida operations, apparently active since prehispanic times. They lie in the immediate vicinity of sites dating from the Classic period on, and were reported to have been in operation at the time of conquest (AGI Guatemala 128, Gall 1968:232). Subsequent reports attest to their continued exploitation throughout the Colonial period (Pardo 1937:288–89; Cortés y Larraz 1958, I:76,82) and nineteenth and twentieth centuries (López 1974:155; DGEYC, *AE* 1944, 1948; Anonymous 1925:50; Lefond 1969:137).

The salinas of Ayacachapa were the last to go, having been exploited until the mid-1970s. In 1978, the area was littered with vestiges of what was once a major active industry: heaps of fragments of large ceramic salt-making ollas, iron peroles (fig. 5.2), mounds of leached earth, as well as abandoned asoleaderos and hornos (fig. 5.3).

Fig. 5.2. Iron perol used for cooking brine in the late nineteenth and early twentieth centuries at Salinas Ayacachapa, Acajutla region, El Salvador

Fig. 5.3. Abandoned asoleaderos (solar evaporation tanks) at Salinas Ayacachapa, Acajutla region, El Salvador. In foreground are collapsed walls of an adobe horno. The iron vat has been removed.

JALTEPEQUE REGION

The Jaltepeque estuary has long been known as a major salt-making region, even though almost all the salinas are presently abandoned. The term "jaltepeque" refers to an earth mound left by sal cocida operations of which many are found around the estuary.

While no prehispanic sites have been reported in the immediate vicinity of the estuary (the area has not been surveyed), there is evidence of saltmaking at the time of conquest (AGI Guatemala 128, Gall 1968) and later during the Colonial period (Cortés y Larraz 1958, 1:145). More recent data clearly indicate that Jaltepeque was a major producer of salt in the nineteenth and early twentieth centuries (Gutiérrez y Ulloa 1962:31, 133; López 1974; DGEYC, *AE* 1944, 1948).

With the exception of two small sal cocida operations near the Lempa River, the salinas of the Jaltepeque region have been abandoned since the early 1960s.

JIQUILISCO REGION

These salinas are not well documented before the nineteenth century. There is a dearth of archaeological data on the region, and only limited evidence of saltmaking in the sixteenth century (AGI Guatemala 128, Chamberlain 1947:641–42). In the nineteenth century, however, a considerable amount of salt was being produced around the Jiquilisco estuary (Gutiérrez y Ulloa 1962:31, 133), which by the 1940s had become the largest salt-producing region of El Salvador (DGEYC, *AE* 1944, 1948).

In 1946 the first solar salt pans were constructed at the Salinas El Aceituno (fig. 5.4), and others followed (Jacobo Handall personal communication 1978). In 1978, the Jiquilisco region was the largest producer in Central America.

Fig. 5.4. Aerial view of Salinas El Aceituno, a modern solar evaporation complex on the Jiquilisco estuary, El Salvador

LA UNION REGION

The salinas of this region lie along the shores of the Gulf of Fonseca near the eastern border of the country. Most of the salt works are located between La Unión and the Honduras border, except for three small sal cocida operations in the vicinity of El Tamarindo farther south (fig. 5.1).

Little is known of the archaeology of the region. Several prehispanic shell middens are located in the area, one of which has remains dating to Late Classic times. Known as Asanyamba, this midden is located in the midst of a modern salt-making locality; its prehispanic occupants may have made salt (Stanley Boggs personal communication 1978).

While there is evidence of saltmaking in this region at the time of conquest (AGI Guatemala 128, Chamberlain 1947), I was unable to locate any accounts of the salt industry during the subsequent Colonial period and nineteenth century. A number of reports on the twentieth-century industry are available (Castro 1905:63; Calderón 1927:88; Cardona Lazo 1939:184–86; DGEYC, *AE* 1944, 1948).

Prior to 1960, the salinas of La Unión were all sal cocida operations. Since then, however, many have been gradually transformed into solar operations, and in 1978 the region was the second largest salt producer in the country.

PRODUCTION AND TRADE

The earliest production figures we have for El Salvador date to 1807 (Gutiérrez y Ulloa 1962:133); at that time production was estimated at close to 2,000 tons a year, which does not include the production of the La Unión area. In 1858 national production was approximately 5,000 tons a year (Andrews n.d.). By the late nineteenth century, El Salvador was not only self-sufficient in salt but was also exporting small quantities to Guatemala (RG, MHYCP, *M* 1883–1900).

In 1944, before the introduction of solar evaporation, the annual production was 13,516 tons. By 1976 that had increased to more than 42,000 tons. Of this amount, almost 20,000 tons were exported to Guatemala, and another 180 tons were exported to Nicaragua, Costa Rica, Belize, and Panama (DGEYC, *AE* 1944–1976).

6.

The Evolution of Salt-Making Technology

THE VARIOUS SALT-MAKING TECHNIQUES of the Maya can be divided into two major categories: sal solar and sal cocida. Ways of employing these techniques vary from region to region, and in many cases, have changed through time.

Many of the techniques date to prehispanic times, while others were developed more recently. Mendizábal (1946a:312-13) has argued that solar evaporation was the oldest technique employed in Mesoamerica, while sal cocida operations were developed much later. This idea has been challenged by Sisson (1973:90) and Dillon (1975:101), and, as the data in this study clearly indicate, it was definitely not the case among the Maya: solar evaporation (in Yucatán) *and* sal cocida techniques (along the Pacific) were both being employed by the Late Formative period, if not earlier.

Mendizábal perceived the primacy of solar evaporation as a cultural phenomenon. In his view, the development of sal cocida technology—which he believed to be superior—coincided with the diffusion of advanced cultures in various parts of Mesoamerica. This is simply not the case. People developed or adopted the technique that would yield the most salt for the least labor in their particular environmental setting. Solar evaporation was the logical choice in Yucatán, but was impossible in pre-twentieth-century coastal Guatemala; thus, the Guatemalans had no alternative to the laborious sal cocida process they employed.

SAL SOLAR

Solar salt technology has been developed in four regions of the Maya area. The sal solar industry of northern Yucatán is the only one that can be assigned safely to prehispanic times; the Pacific coast industry of Guatemala and El Salvador was developed during the present century, and the origins of the Chiapas solar salt-making techniques have not been documented.

Yucatán

Almost certainly people were making salt in northern Yucatán by the early phase of the Late Formative period, from about 300 to 50 B.C. Several colleagues working in Yucatán would concur with me that saltmaking began at an earlier date; ongoing investigations suggest that this may be the case (Andrews V et al. 1981).

Early saltmaking in Yucatán was probably a simple process, consisting merely of collecting the salt that formed naturally on the shores of dry lagoons and estuary coves. Such activities are not visible in the archaeological record, which is why I have doubts about uncovering evidence of any saltmaking before the Late Formative. In fact, our earliest evidence of saltmaking is not based on actual salt works, but on the fact that Late Formative campsites and hamlets are located amid ideal saltmaking areas. In addition, Eaton (1974) has reported that shell "celts" are commonly found at these sites; I believe that many of these tools served as salt scrapers, a use that survived into the twentieth century (Hester 1953:292).

It is not clear when people began building actual salt works. The oldest known salt pans and storage platforms date to the end of the Late Formative period, and they became widespread during the Early Classic period. From that time on, the industry's technology is remarkably continuous. To this day, many salinas of the north coast are exploited much as they were during the Early Classic period.

Prior to the mid-twentieth century modernization at Las Coloradas, the only factor that would have limited salt production was transportation. The prehispanic Maya had only canoes and bearers, while the Spanish had ocean-going ships and horse-drawn carts; further development of transport facilities came in the early part of this century with the appearance of trucks and improved roads and has played a major role in the growth of the industry.

Guatemala and El Salvador

Solar salt industries began in Guatemala in the 1920s and in El Salvador in the 1940s. These modern enterprises, which require major long-term capital investments for construction and maintenance of complexes of asoleaderos and salt pans, represent a clear break with the traditional cottage industries.

There is little variation in the technology of this industry. The only feature that has changed is the solar patio, or salt pan. In the early years in Guatemala saltmakers experimented with wooden and concrete patios, eventually switching to tile, which most now use. In eastern El Salvador I observed one ingenious method of avoiding the expense of the tile. The saltmakers had built earthen patios and then covered them with huge sheets of black plastic. I was told this was a makeshift arrangement because they could not yet afford permanent tile patios.

The sal solar industries of Guatemala and El Salvador now produce more than 80 percent of their total national production. If present patterns continue, they will eventually replace all sal cocida operations and double their production by the year 2000.

Chiapas Coast (Tapesco System)

Unfortunately, we have scant historical information on this industry, and thus are ignorant of its antiquity. It is possible that this industry is a recent development which superseded an older sal cocida industry; on the other hand, the tapesco method may date to colonial and possibly prehispanic times. Future historic research may solve this problem.

Chiapas Highlands (La Concordia)

The now-extinct salt industry of La Concordia is unique in the Maya area, for it is the only locality in the highlands where solar evaporation is known to have been used.

As in the case of the tapesco method utilized on the coast, we have no idea of the antiquity of the solar evaporation method of La Concordia: it may be relatively recent—nineteenth century—and in earlier times the salt may have been acquired by a sal cocida process. I find this difficult to accept, however, for the technique employed at La Concordia was employed elsewhere in Mesoamerica in prehispanic times.

In the Tehuacán Valley, in the highlands of central Mexico, a similar process of solar evaporation is used. This industry is long-lived: archaeological excavations have uncovered solar salt pans dating to Postclassic times (Sisson 1973). In view of this, it is quite possible that the saltmakers of La Concordia were also using solar evaporation in prehispanic times.

SAL COCIDA

The sal cocida technology of the Maya area is far more diverse than that of solar salt and has changed considerably during the last 150 years. Most of these changes have taken place on the Pacific coast of Guatemala and El Salvador.

Pacific Coast

The sal cocida industry of the Pacific coast of the Maya area is possibly one of the oldest in Mesoamerica. On the coast of Chiapas, saltmaking may have begun in Early or Middle Formative times, and by the Late Formative period it was probably common all along the coast, from Tehuantepec to the Gulf of Fonseca.

Our knowledge of the prehispanic salt industry is scant, since its archaeological visibility is low. Abandoned prehispanic or Colonial period sal cocida operations offer only a few potsherds strewn among mounds of leached earth. The mounds, of course, soon became part of the landscape; I doubt that any of the leached-earth mounds I saw during my survey were more than 200 years old.

Also found at prehispanic sal cocida sites are the ceramic objects known as "salt cylinders." These artifacts, believed to have served as props for the slow cooking of brine, have been found at salt-making sites on the coasts of Guatemala and Costa Rica (Coe and Flannery 1967:91, pl.28; Edwin Shook personal communication 1978), at Sacapulas in highland Guatemala (personal observation), and in the Tehuacán Valley in central Mexico (Sisson 1973). With the exception of Sacapulas, where stone cylinders are still used, all the known clay cylinders are prehispanic, ranging from Late Formative (Ocós and Salinas Acapán) to Postclassic times (Sacapulas and Tehuacán Valley). Although I was unable to document the use of these clay cylinders in either colonial or modern times, I suspect they were no longer in use in the Colonial period along the Pacific coast, as several early accounts of the sal cocida industry allude to the practice of placing brine-cooking vessels over adobe ovens rather than cylinders (García de Palacio 1927:72; Fuentes y Guzmán 1932–33, II:102–3). Similar clay cylinders have also been reported from prehistoric salt-making sites in Europe (Riehm 1961).

The sal cocida industry of the Pacific coast appears to have remained unchanged throughout the Colonial period. Beginning in the mid-nineteenth century, many new technological features appeared—iron peroles, iron vat ovens, asoleaderos—and production began to increase. The culmination of this process was the adoption of solar evaporation, which today has become the predominant technology. It is unlikely that the remaining sal cocida operations will survive beyond this century.

Inland Sources

The sal cocida salt works of Chiapas and Guatemala can be divided into highland brine springs and salt playas, and the single lowland brine spring and salt playa of Salinas de los Nueve Cerros. Most of the inland sources appear to have been exploited in prehispanic and colonial times. Only four were still in use in 1978: Ixtapa and Atzam in Chiapas, and San Mateo Ixtatán and Sacapulas in Guatemala.

Prior to the mid-nineteenth century, brine was cooked in ceramic vessels at all of the inland salinas. Thereafter, however, several localities switched to iron vat ovens: Ixtapa, Salinas Magdalena, Salinas de los Nueve Cerros, and the salinas around Barillas. With the exception of Ixtapa, all of these salinas have been abandoned. The other remaining salinas that are still being exploited—Atzam, Sacapulas and San Mateo Ixtatán—employ techniques that have not changed since prehispanic times.

SEASONALITY AND SPECIALIZATION

With the exception of a few salinas in the highlands, saltmaking in the Maya area has traditionally been a seasonal occupation: most salt is produced during the dry season, from January to June. In northern Yucatán, only maintenance work is required during the early months, and intensive labor is required only during May and June, when the salt is harvested.

The seasonality of saltmaking is convenient to most Maya salt-makers, who are generally engaged in agricultural work the rest of the year: because the lull in farmwork occurs during the peak of the salt-making season, thousands of farmers can go to the coasts to earn supplemental income.

Formerly, this scheduling was important to scores of communities located within walking distance of the coast. Groups of men from inland regions would travel to the coast for a few weeks to gather a year's supply of salt for their villages. Since they did this during the dry season, their absence did not disrupt planting or harvesting.

Highland salt-making communities were not so affected by seasonality, since most of the salinas could be exploited year-round. Such is the case of towns with salt springs, like Ixtapa, Atzam, and San Mateo Ixtatán.

Some inland salinas, which were exploited with the aid of the sun, were affected by the weather. La Concordia was exploited only during the dry season. At Sacapulas the extraction of salty soil from the playas could only be carried out during the dry months. This soil could, however, be stored in sheds and processed after the rains began. I believe the same was true at Amatitlán, where the salt-making method was similar to that of Sacapulas.

Finally, the salt flats of Salinas de los Nueve Cerros may have been exploited by solar evaporation of brine prior to cooking. If such was the case, it could have only been done for short periods of time, as Nueve Cerros has one of the wettest climates of the tropical lowlands, with almost year-round precipitation (Dillon 1977:4).

Several colleagues have suggested that the seasonality of the salt industry would have affected the scheduling of the salt trade. While it is likely that the volume of salt in the trade networks increased during major production periods (i.e. spring and summer), the trade was by no means limited by the seasonality of production.

Salt can be and was stored for extended periods of time. Along the Pacific coast of the Maya area, it was placed in thatch sheds, whereas in Yucatán it was collected in large piles which were either covered with palm thatch or glazed with torches. Thus, it could be shipped to market whenever the producer wished. In short, the trade probably obeyed the laws of supply and demand rather than the vagaries of seasonal climate.

However, one aspect of the trade that may have been affected by seasonal factors was river transport. Many rivers of the Maya area rise considerably during the rainy season, permitting boatmen to go farther upriver. This may have been an important factor in the salt trade—salt is heavy and large quantities were being transported. Had the Maya waited until the rivers rose to their highest levels in late summer and early fall, they could have significantly reduced the overland transport distances.

A final consideration is the nature of saltmaking as a specialization. Among the Aztecs, the production and trade of salt was a highly specialized industry involving entire communities. The saltmakers formed guilds and held formal festivities every year in honor of Huixtocihuatl, the goddess of salt. Aztecs involved in the salt industry were full-time specialists dedicated to the various ramifications of production and commerce of their product (Noguera 1975).

By contrast, the Maya salt industry does not appear to have been as highly specialized or institutionalized. Moreover, because of the seasonal nature of saltmaking in most areas, Maya saltworkers were not full-time specialists. In fact, most spent the greater part of the year tending their fields and in other pursuits. The only people who could be seriously called salt specialists are the people—mostly women—who cook brine all year in highland salt-making communities. While these women tend to a variety of domestic chores, they also invest great time and effort in making salt, which once represented a major source of income for their villages.

Another interesting feature of the traditional Maya salt industry is that in many localities the salt trade was handled by merchants who came from communities other than those producing the salt. The salt from Ixtapa, for example, has traditionally been traded and marketed by Zincanteco merchants; likewise, the salt of San Mateo Ixtatán is often traded by itinerant merchants from Santa Eulalia, San Juan Ixcoy, and Soloma, a pattern that dates to the Colonial period (Olaverreta 1935:19). Much of the salt of La Concordia was purchased by merchants from Comitán, which is the main reason it was referred to as Sal de Comitán (Manó 1940:511, Blom 1945:4). In the small traditional salt works along the Yucatecan and Pacific coasts, I observed that the salt was rarely marketed by the producers; rather, it was purchased by middlemen who then marketed it.

That a large portion of the native salt trade was handled by specialized merchants is not surprising. Such a pattern dates to prehispanic times throughout the Maya area (Roys 1943; Feldman 1971). In Yucatán the most prominent long-distance traders were noblemen who capitalized on the rich production of the north coast beds; that production, and their trading patterns, had an enormous impact on the development of Maya civilization. To appreciate how this came to be, one needs a diachronic view of salt production and trade in the Maya area.

7.

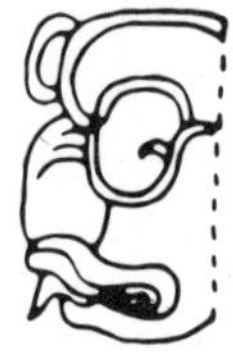

History of Production and Trade

AT THE TIME OF THE SPANISH CONQUEST the Maya salt trade consisted of two major networks: one was based in northern Yucatán and supplied local markets as well as distant populations throughout the Maya lowlands. The other was the regional trade network of the highlands and Pacific coastal plain, which supplied most communities with salt from nearby salinas. The Yucatecan trade involved large quantities transported over long distances, often over water. By contrast, the southern trade was land-based and involved transportation of small quantities over short distances.

This general reconstruction of the Maya salt trade, based on extensive historic and ethnographic data, can be projected into the prehispanic past. Such a projection, however, need not be viewed as a simple ethnographic analogy, because it is buttressed by several centuries of well-documented historic continuity.

The model is further reinforced by a global analysis of the quantitative aspects of the Maya salt industry. The key to this quantitative approach lies in the balance between production and consumption. Simply stated, it involves determining where a given population acquired the salt it needed in order to survive.

To satisfy the minimal salt needs of some five million people (the estimated population of the Maya lowlands in the Late Classic period) would require approximately 15,000 tons of salt a year. The logical question, then, is where this salt was acquired. The primary objective of the survey of salt sources was to answer this question by gathering data on production.

PRODUCTION

A regional breakdown of prehispanic and modern production is presented in Table 7.1 summarizing data presented in previous chapters.

I believe the chart answers the basic question regarding the provenience of the salt consumed by the lowland Maya during the Late Classic: Yucatán was the only producer that could have met the demand of the lowland population.

The production of other regions was considerably lower and was likely consumed by local populations. Chiapas, for example, had a production potential of 2,300 tons, enough to meet the basic needs

Table 7.1
Prehispanic and Modern Salt Production in the Maya Area
(metric tons per annum)

Region	Prehispanic Production*	1980 (est.)†
Yucatán		
All sources	**20,000**	**430,000**
Chiapas		
Central Highlands	**100**	**35**
La Concordia	**200**	—
Pacific Coast	**2,000**	**300**
Guatemala		
San Mateo Ixtatán	**100**	**50**
Sacapulas	**70**	**10**
Other highland sources	**100**	—
Salinas de los Nueve Cerros	**300**	—
Pacific Coast	**5,000**	**30,000**
El Salvador		
Pacific Coast	**5,000**	**42,000**
Totals	**32,870**	**502,395**

*The sources for these figures are cited in Chapters 2 through 5. They are maximum potential production figures, given prehispanic salt-making technology; the exception is Yucatán, which had the potential to produce much larger quantities, perhaps as much as 60,000 tons.

†Modern production figures reported in Chapters 2 through 5 provide the basis for these estimates. The figure for Yucatán was provided by José Palomeque of the Industria Salinera de Yucatán, S.A. (personal communication 1980). Regions without modern production figures are those in which the salinas have been abandoned.

of approximately 787,670 people. I suspect, however, that Chiapas had a higher population during Classic and Postclassic times, and therefore must have imported salt from elsewhere, namely Tehuantepec and Yucatán.

The salinas of the highlands and Pacific coast of Guatemala had a total production potential of 5,270 tons, enough to supply 1.8 million people. Though concrete figures are not available, it is doubtful that the prehispanic (Late Postclassic) population of highland and Pacific coast Guatemala was less than 1.5 million; in fact, it was probably higher (Thompson 1966:29, Robert Carmack personal communication 1977). In sum, this region had little or no salt to spare, and may have been importing salt in prehispanic times. Thompson (1970:149–50) has suggested that Yucatecan salt was imported into the eastern highlands via Lake Izabal, and both he (1970:135) and Feldman (1978:14–15) have implied that the salt from Nueve Cerros also reached the highlands. Since highland Guatemala imported

British and Salvadorean salt during the nineteenth and twentieth centuries, it is likely that it also imported salt in prehispanic times.

Finally, there is no evidence to suggest that El Salvador exported any salt to other parts of the Maya area prior to the late nineteenth century.

PREHISPANIC TRADING SPHERES

The Maya salt trade networks operated in two almost mutually exclusive spheres. The concept of trading spheres has not been formally defined and should not be confused with the concept of sphere employed in ceramic typology (Willey et al. 1967:306), though it too is useful in reconstructing ancient trade. A trading sphere is simply the trade range of a given commodity from an established place of production. This concept is similar to the notion of "regional exchange networks" of particular commodities described by Pires-Ferreira (1975). To date, trade range studies have focused on such items as obsidian, iron-ore mirrors, shells (Pires-Ferreira 1975), Mayan lithic artifacts (Rovner 1975), and volcanic ash temper (Simmons and Brem 1979).

In the case of salt, the northern trade sphere included the entire Maya lowlands from Tabasco to Honduras. Throughout most of their history, the people living in the northern sphere acquired their salt from the coastal lagoons of northern Yucatán, and, to a lesser degree, from the minor sources located along the Gulf and Caribbean coasts and Salinas de los Nueve Cerros. The salinas at Nueve Cerros were likely a critical source for the populations of the upper Usumacinta and Pasión river drainages. Given the geographic distribution of the salinas, large areas were supplied through long-distance trade networks. The bulk of the trade moved along the coasts and up rivers. From river terminals it continued by land to destinations in the interior (fig. 7.1).

The salt trade of the southern sphere appears to have been quite different. Because highland salt springs and coastal salinas were dispersed throughout the southern sphere, most communities were located less than 120 km from salt sources. Thus, distribution of salt was handled through highly localized trade networks involving short distances and small quantities. Even though most highland and Pacific coast salinas had a low production potential, collectively they were able to supply most of the needs of the population living in the southern sphere.

Salt may have crossed from one sphere to another in only a few, apparently minor instances. Salt from Yucatán and Salinas de los Nueve Cerros may have reached the highlands, but this has yet to be documented. Also, the salinas of La Concordia may have supplied small quantities of salt to the adjacent Chiapas lowlands.

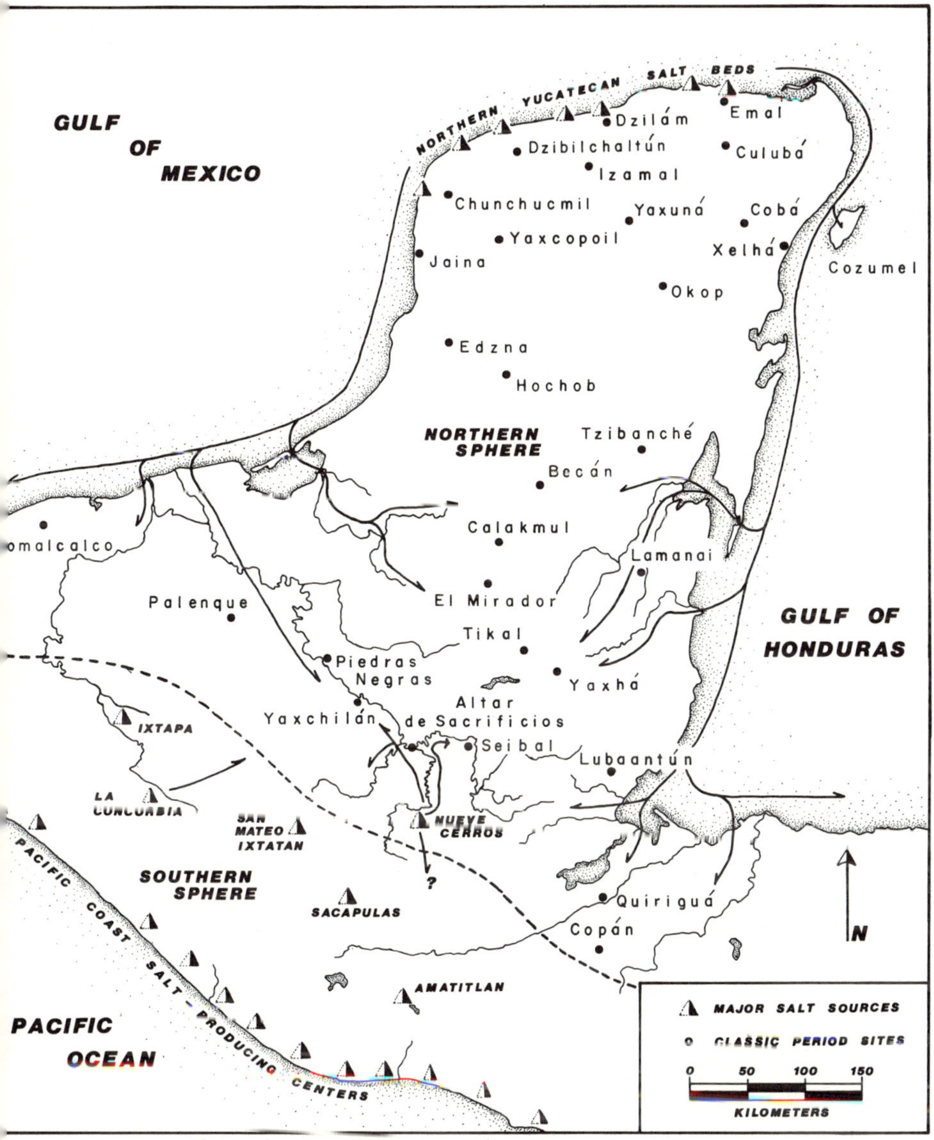

Fig. 7.1. Salt-trading spheres of the Maya area and major salt trade routes of the Classic period, ca. A.D. 300 to 1000

The two major trading spheres can be broken down into several subspheres, although they are ill defined. Four subspheres are evident in the north: (1) the internal overland trade of northern Yucatán; (2) the western trade of the Gulf Coast–Usumacinta drainage (often referred to historically as the Campeche salt trade);

(3) the eastern trade of the Caribbean; and (4) the trade range of the Salinas de los Nueve Cerros. I say these are not well defined because the total extent of the trade range of each subsphere, and their shifts and overlaps through time, are not known. For example, we might ask where the inhabitants of Becán, a large Classic period site located at the center of the base of the peninsula, acquired their salt: from overland trade from the north, or from the eastern or western trade routes? The trade ranges of salt from Yucatán and Salinas de los Nueve Cerros exemplify overlapping subspheres. I suspect that salt from both sources was available in the same areas along the middle and upper Usumacinta drainage during the Classic period.

In the southern sphere every salt-making locality had a distinct trade range, and the entire region might be viewed as a cluster of small regional subspheres. These, too, considerably overlapped and likely changed through time. For instance, a visitor to the Sunday market in Huehuetenango in the late nineteenth century may have encountered salt vendors from Comitán (La Concordia), San Mateo Ixtatán, Sacapulas and the coast, all selling side by side. Changing political structures may also have brought about changes in the trading networks. For example, the Cakchiquel revolt against the Quiché in the fifteenth century may have led to a reduction of shipments of coastal salt into the Quiché realm, which in turn would have created a greater demand for the salt from Sacapulas.

DEVELOPMENT OF SALT TRADE NETWORKS

We have good reason to believe that the production and trade of salt began during the Formative period, but because of demographic factors and the geographic distribution of the mineral the evolution was different in the various regions of the Maya world. In many places the trade never went beyond a small-scale exchange, like that seen today in the highlands of Chiapas and Guatemala. In other areas, however, the demand for salt generated extensive trade networks whose logistics underscore the complexity of the prehispanic economic operations.

Salt was the heaviest and bulkiest of the long-distance trade goods exchanged by the Maya, and all available evidence indicates that it was moved by water whenever possible, an obviously ideal mode of transport (Roys 1943:52; Scholes and Roys 1948:59, 244; Thompson 1970:149–50). Even by water, however, the logistics of the long-distance trade are staggering. During the Classic period, when the Petén imported thousands of tons of Yucatecan salt each year, transport undoubtedly required large fleets of canoes. Numerous slaves must have served as paddlers, with bearers to carry the salt from the river terminals to destinations in the interior, where many of them probably were sold along with the salt, a common practice among traders at the time of conquest (Roys 1943:53). All told, the trade involved hundreds, and perhaps thousands of people.

Given the sheer volume, complex logistics, and critical nature of this trade, it likely was administered by highly organized groups of merchants who accrued great wealth. Such was the situation at the time of conquest when the trade was in the hands of the nobility (Roys 1943:51); it is quite possible that this was also the case during the Classic period.

If so, one can surmise that these traders were powerful people, for in controlling the one resource vital to the population of the southern lowlands, they also controlled the coastal trade networks between the northern and southern networks. A comparable situation developed in West Africa during the Middle Ages when Moroccan merchants, who exploited the salt works of the Sahara, also dominated the trade networks between North Africa and the Sudan (Bovill 1968). For the Maya area the ramifications of such control are overwhelming: whoever controlled the salt trade played an important role in the development of lowland Maya civilization.

While ample evidence indicates that the growth of the salt industry paralleled the emergence of social stratification, urban centers, and states, chronological links between these processes are not easily demonstrated. One of the most difficult objectives of the Maya salt project was establishing a time framework for the development of salt-making industries and salt-trading networks. My initial expectations were not high, as the original research design envisioned only a survey of the salinas and a synthesis of the available archaeological, historic, and ethnographic data on each salt-making locality. An ideal extension of this research would have been a vertical excavation program at each salt-making site, but time and funding were not available for such an endeavor. Consequently, the prehispanic record of exploitation of any given salina must remain as an objective of future excavation projects.

Nonetheless, I was able to gather a fair amount of chronological data from many salinas. The resulting picture, although limited, far exceeded my first hopes: these data are summarized in Table 7.2, which presents a synopsis of the known history of exploitation of each salt-making locality discussed here. Despite certain gaps, the data do permit a preliminary reconstruction of the history of the salt industry and its repercussions on Maya culture.

FORMATIVE PERIOD

The harvest of naturally occurring salt along the north coast of Yucatán probably began early in the Formative period, but it is unlikely that we will ever find any archaeological evidence of these elusive beginnings.

Table 7.2
Chronology of Exploitation of Salt Sources In the Maya Area and Gulf Coast Region
• = definite evidence of exploitation
? = uncertain evidence of exploitation

Salt Sources	Late Formative (300 B.C.–300 A.D.)	Early Classic (300–600 A.D.)	Late and Terminal Classic (600–1000 A.D.)	Postclassic (1000–1500 A.D.)	16th Century	17th & 18th Centuries	19th Century	20th Century	Active in 1978
Yucatecan beds	•	•	•	•	•	•	•	•	•
Holbox, Quintana Roo								•	
Isla Mujeres, Quintana Roo				•	•		•	•	
Cozumel, Quintana Roo				?	?				
Ambergris Cay, Belize				•			•	•	
Isla del Cármen, Campeche							•		
Acalapa, Veracruz				•	•				
San Andrés, Veracruz	?	?	•	•	•	?	•	•	
Tlacotalpan, Veracruz				•	•				
Quiahuistlán, Veracruz				?	?		•	•	
Pánuco sources, Veracruz				•	•	•			
Tamaulipas coast						•	•	•	•
Chiapas coast	?	?	?	•	•	•	•	•	•
Ixtapa, Chiapas	•	•	•	•	•	•	•	•	•
Atzam, Chiapas				•	?	?	•	•	•

The oldest evidence of saltmaking in Mesoamerica dates to the Early Formative period in the Valley of Oaxaca (Drennan 1976). Although the earliest evidence is tentative, saltmaking in the Maya area may also date to that period on the Pacific coast of Chiapas (Paillés 1978). Another possible early salt-making locality is Salinas de Portatenco, near La Concordia, Chiapas, where archaeological

Table 7.2
Chronology of Exploitation of Salt Sources In the Maya Area and Gulf Coast Region
(continued)

Salt Sources	Late Formative (300 B.C.–300 A.D.)	Early Classic (300–600 A.D.)	Late and Terminal Classic (600–1000 A.D.)	Postclassic (1000–1500 A.D.)	16th Century	17th & 18th Centuries	19th Century	20th Century	Active in 1978
La Concordia, Chiapas	•	?	?	?			•	•	
Guatemala coast	•	•	•	•	•	•	•	•	•
San Mateo Ixtatán, Huehuetenango			?	•	•	•	•	•	•
San Miguel Acatán, Huehuetenango			?	?			•	•	
Barillas, Huehuetenango							•	•	
Las Victorias, Quiché							•		
Salinas Magdalena, Quiché			•	•	•	?	•		
Sacapulas, Quiché			?	•	•	•	•	•	•
Sabalpop, Chimaltenango							•	•	
Amatitlán, Guatemala		•	•	•	•	•	•	•	
Nueve Cerros, Alta Verapaz	?	•	•	?	•	•	•	•	
El Salvador coast	?	?	•	•	•	•	•	•	•

remains directly above the salt works date to Middle Formative times (Warren 1978).

By Late Formative times, saltmaking was underway throughout the Maya area, most notably along the coasts of Yucatán and Guatemala. Although sufficient evidence is lacking, I believe that future excavations will confirm that almost every salt-making locality of the Maya region was being exploited by the beginning of the Christian era.

The basic reason behind this belief is the sizeable and fairly dense population of the Maya area at that time. Despite a lack of population figures, investigations have revealed that Maya communities were much larger in mid- to Late Formative times than originally believed. In the lowlands, communities such as Komchén, Cobá, Cerros, Lamanai, Becán, Uaxactún, Tikal, Edzná, Mirador, Altar de Sacrificios, and Seibal had substantial numbers of inhabitants. Formative sites are less well known in the highlands, but we now know that the Central Depression of Chiapas and the Valley of Guatemala supported large populations in Late Formative times. Finally, excavations have revealed a fair number of Formative communities along the Pacific coast, particularly in the Iztapa-Altamira region of Chiapas, and in the vicinity of Ocós, Champerico, and Puerto San José in Guatemala. Moreover, the vast majority of excavated Maya sites have been shown to have been occupied in Late Formative times.

A critical consideration in determining the origins and development of the long-distance salt trade in the Maya area is the nature of the diet of Formative times. Many scholars have argued that the origins of saltmaking and salt trading can be traced to the neolithic revolution, when a shift to a low sodium diet led to a demand for salt (cf. Nenquin 1961, Bloch 1963).

In the Maya area the spread of agriculture began more than 4,000 years ago, but the shift to a predominantly cereal diet undoubtedly lasted centuries. Early Maya farmers were also hunters and were probably able to satisfy their sodium needs with an abundance of game.

However, several centuries of population growth, deforestation, and hunting likely led to a depletion of faunal resources. Such a process has occurred in northern Yucatán during the last century. The depletion of faunal resources was, then, the final stage in the shift to a low sodium diet consisting of cereals, vegetables, and fruit. With this shift, salt became a necessity and its production and trade became an important facet of Maya life.

We may never be able to determine when this change occurred in the Maya area, but available evidence suggests that it happened sometime during the Late Formative period. The significant increase in population may have given rise to land shortages (Ball 1977a, Webster 1977, Sanders 1977). Recent pollen data from the central Petén support this notion: it appears that by the end of the Formative period this area had been largely deforested (Deevey et al. 1979). In addition, a growing amount of osteological data suggests that the Maya diet from Late Formative times on included little meat.

The appearance of a widespread salt industry in northern Yucatán in Late Formative times as well as coastal outposts which would have monitored and assisted in the logistics of salt transport is further evidence of the existence of a large-scale trade. The most prominent coastal outpost is the large Late Formative site of Cerros, in Chetumal Bay at the entrance of the New River, a major access route into the southern lowlands. Cerros undoubtedly owed its existence to the maritime trade between north and south, and salt was most likely one of the items involved in that trade (Freidel 1978, 1979).

While it is arguable that a Late Formative long-distance salt trade arose to satisfy nutritional needs, it is doubtful that changing dietary patterns constitute a sufficient cause. Although I regard the neolithic revolution, then, as a convenient *deus ex machina* to explain the existence of salt-trading activities, I suspect the origins of the trade are as readily traced to the human palate as to dietary necessity. After all, salt is a popular condiment, and the Maya may have been trading it long before it became a critical part of their diet.

CLASSIC PERIOD

Northern Sphere

During the Early Classic period, the Maya salt trade developed into a major industrial and commercial operation. The shift on the north coast of Yucatán from small salt-making hamlets to larger operations capable of producing salt in industrial quantities was probably a response to a growing population and resulting increased demand for salt throughout the Maya lowlands. Evidence that clearly supports the emergence of a regular north-south trade is the appearance of substantial quantities of obsidian and Early Classic polychrome ceramics in northern Yucatán. As Andrews IV (1969:60) has suggested, these exotic items from the south were probably acquired in exchange for salt. The appearance of these trade items at north coast salt-making sites only reinforces such a notion.

One aspect of the western salt trade of northern Yucatán not presently understood is how far it extended around the Gulf coast during the Classic period. I have few doubts that it reached past the shores of Tabasco into southern Veracruz, as these areas were not self-sufficient in salt. Data support the notion that there was a fair amount of intercourse both in goods and ideas between Yucatán, the southern Gulf coast, and the central Mexican highlands during the Classic period (Thompson 1953; Andrews V 1974, 1979; Ball 1977a, 1977b, 1978; Nelson 1980). Much of this evidence consists of goods (green obsidian) and cultural influences, (in architectural style) that emanated from central Mexico and reached Yucatán through intermediary groups of the southern Gulf coast. By acting as middlemen, these groups were able to acquire salt from northern Yucatán. I doubt that the Yucatecan salt went farther than the Gulf coast lowlands. Ball (1977a:182) has suggested that Yucatecan salt was traded to the central Mexican highlands in prehispanic times, but I have found no evidence of such a trade. In fact, the existence of large-scale prehispanic salt industries in the valleys of Mexico and Puebla would argue against it.

The economic impact of the Yucatecan salt trade was substantial and was probably a major factor in the rise of urbanism in northern Yucatán (Andrews IV 1968; Andrews V 1974; Ball 1977a, 1977b). As several scholars have noted, during the Late Formative and Classic periods a series of major communities emerged in the immediate interior of the north coast of Yucatán. Located anywhere from 20 to 30 km inland, at the edge of marginally arable land, these communities were spaced at intervals along the coast: among them are Chunchucmil, Tzeme, Dzibilchaltún, Dzilám González, and San Fernando. These sites supported fairly large populations (10,000 to 40,000 people) (Kurjack personal communication 1979), which far exceeded the agricultural carrying capacity of their surrounding regions. Thus it is likely that the inhabitants of these communities were involved in nonagricultural pursuits such as exploitation of marine resources, craft production, and saltmaking (Vlcek et al. 1978).

When the Spaniards arrived in Yucatán, they found the north coast under the control of different provinces whose capitals were in the interior (Roys 1957). Such a pattern may have also been the case during the Classic period. The Classic period centers probably each controlled a stretch of nearby coastline whose marine resources and saltbeds they exploited. While direct associations are difficult to make on the basis of available data, it is probable that Chunchucmil and Tzeme controlled the Celestún beds, Dzibilchaltún (and neighboring Komchén) the Chuburná-Progreso-Chicxulub beds, Dzilám González the Mina de Oro beds, and San Fernando the Emal beds.

This model, while admittedly tentative, would partially explain the raison d'être of these large communities. Control of the coast would have allowed them to exchange salt and other marine resources for agricultural foodstuffs from the interior, a pattern that would explain the ability of these centers to support large nonfarming populations. But the factor most likely responsible for the growth and accumulated wealth of these communities was their participation in the long-distance coastal trade networks that girded the peninsula. Their control of salt production gave them an inordinate amount of power, if not outright control of these networks.

The salt trade probably had a similar impact on the communities that controlled the other "end" of the long-distance trade networks, particularly those in the central area of the southern lowlands. Rathje's (1971) core–buffer zone theory postulates that the need to acquire certain critical resources, of which salt was one, promoted the organizational impetus that gave rise to the large communities dominating the core area during the Classic period. While parts of Rathje's theory are debatable, I believe the data in this study clearly reinforce the notion that the salt trade was a powerful ingredient in the development of Classic lowland Maya civilization.

Southern Sphere

The regional salt trade networks of the southern sphere probably originated in the Late Formative period and were well established by the Classic period. We can only speculate on the effect that the rise of Kaminaljuyú had on the salt industry. The only source of salt in the Valley of Guatemala was that of Amatitlán, and I doubt that it was able to produce quantities sufficient to satisfy the needs of the valley population during the Classic period. Thus, it is likely that the valley acquired additional salt from the nearby Pacific coast, giving rise to a large industry in the San José–Iztapa area. This region, the major salt producer when the Spanish arrived, continues to be so this day.

TERMINAL CLASSIC AND POSTCLASSIC PERIODS

Northern Sphere

This was a time of considerable turmoil and change in the lowlands, beginning with the collapse of the southern cities and culminating with the Spanish conquest of northern Yucatán in 1542.

The processes that led to the southern collapse began in earlier times and are too complex and numerous to discuss here (cf. Culbert 1973, 1977; Sharer 1977). As several scholars have argued, many of the factors that gave rise to Classic Maya civilization ultimately helped shape its demise. In addition to internal weaknesses, the Maya of the core area faced external economic competition and possibly even outside military pressure.

The external economic competition is now viewed as a crucial component of the collapse. Stated simply, it is believed that the Classic centers of the core area became isolated from the trade networks of the buffer zone, which were in the hands of increasingly powerful maritime merchant groups who bypassed the land-locked core area.

Christopher Jones (1979) has argued that the city of Tikal emerged as a major trading center, a middleman in the east-west river trade of the base of the peninsula; its eventual demise came with the shift to maritime trade which bypassed the Petén. I would add to this argument that Tikal was a middleman in north-south trading networks as well: two of the major resources moving through Tikal were jade and obsidian, which came from the highlands and the Motagua River basin to the south.

Southern lowland cities such as Tikal undoubtedly acquired these resources in exchange for lowland products: cacao, cotton, polychrome pottery, exotic pelage and plumage, tobacco, wood craft objects, and so on. Only part of the jade and obsidian acquired from the highlands remained in the core area; the balance was most probably traded farther to the north, undoubtedly in exchange for salt and other northern products. In short, then, it is likely that *the core area used highland resources to acquire Yucatecan salt.* They may have also exported other items and ideas (Rathje 1977) to the north, but I suspect the Yucatec Maya were primarily interested in the basic materials, namely jade and obsidian.

To return to Jones's hypothesis, when the long-range trade networks began bypassing the core area, cities such as Tikal lost their middleman status and faced a serious economic dilemma. In fact, it is impossible not to wonder if the core was able to acquire significant quantities of salt once the trade routes began shifting. By no means do I wish to imply that the Classic Maya collapse was brought on by a breakdown in the salt trade. Rather, I propose that the potential effects of such a breakdown be included in the many ills that plagued the final decades of southern lowland civilization.

As Jones has noted, the bypassing of the core has an excellent analogy in the trans-Saharan trade of the Middle Ages, in which Moorish traders exchanged Saharan salt for gold from the Sudan; this gold was carried across the Sahara to North Africa and then traded into the Mediterranean and European markets. When European traders began circumnavigating West Africa in the fifteenth and sixteenth centuries, they effectively bypassed the Saharan trade of the Moors, which consequently fell into decline.

That the trading routes did eventually shift is now reasonably well documented. Trace analysis of obsidian samples indicates that most of the Late Classic obsidian in the southern and central lowlands came from El Chayal, probably over land routes; moreover, Classic and Early Postclassic period samples from Cobá, in eastern Yucatán, also came from El Chayal. These probably went overland to Tikal and then by water up the east coast. In contrast, Late Postclassic samples from eastern Yucatán (Cancún, Cozumel) came almost exclusively from Ixtepeque, a source east of El Chayal; this obsidian probably traveled down the Motagua and up the east coast, bypassing the entire core area (Nelson 1980; Phillips 1979: Appendix B).

What happened along the east coast during the Terminal Classic and Early Postclassic periods is not entirely clear, but I suspect that the declining population of the core area brought about a decline in the salt trade.

By contrast, the western salt trade appears to have flourished despite the collapse of the core. During the Terminal Classic period, trading activities between the southern Gulf coast and western Yucatán intensified. This situation appears to have been a result of growing aggressiveness on the part of southern Gulf coast peoples who had acted as intermediaries between the Maya and central Mexico in earlier times. These groups spread down the Usumacinta drainage as far as Seibal in the ninth century (Sabloff and Willey 1967) and by the tenth century had deployed themselves along the west and north coasts of Yucatán (Andrews 1978b).

During this period of expansion, these people, variously known as Toltecs, Chontal Maya, or Putún, were a powerful medium for economic and cultural exchange. Central Mexican sculptural traits spread down the Usumacinta (to the site of Seibal) and up into northwestern Yucatán (to Puuc cities). At the same time, Maya artistic elements began appearing at sites like Cacaxtla and Xochicalco in the central Mexican highlands.

These trading groups appear to have gradually taken over the coastal and major riverine trade networks of the western lowlands, and eventually emerged as a powerful merchant class capable of shaping future events throughout the Maya area (Sabloff and Rathje 1975b).

In the course of their expansion, these merchants must have established alliances with ruling elites in both Mexico and the Maya area. Indeed, some historic evidence suggests that such was the case (John Molloy personal communication 1980). One likely affiliation was with the elite of the cities of the Puuc region of northern Yucatán, where Gulf coast ceramics (Balancan "Z" Fine Orange) and central Mexican sculptural traits are commonly found. Such an alliance would explain the diffusion of certain Yucatecan cultural elements to other parts of the Maya area: Silho "X" fine orange ceramics and Puuc-like architectural traits have been found as far south as Seibal in the southern lowlands (Str. A-3; cf. Willey et al. 1975). This site, and nearby Altar de Sacrificios, had come under the control of the Gulf coast merchant groups; they may have also controlled the salt works at Salinas de los Nueve Cerros (Sabloff and Willey 1967; Thompson 1970; Sabloff 1971; Adams 1971, 1973).

The Gulf coast merchant groups undoubtedly spread down the Usumacinta drainage in search of resources and markets, but in the end they faced a dying civilization that offered few prospects for an aggressive trading people. Ironically, their own economic and military aggressiveness may have hastened the collapse. The southern collapse ultimately led the merchant groups to focus their attention elsewhere. The logical choice was Yucatán, for it was growing increasingly vulnerable to outside pressures, and the merchant groups already had a foothold along its coasts.

During the Terminal Classic period, northern Yucatán began to show symptoms similar to those which caused the southern collapse, the most apparent being overpopulation and land shortages.

Archaeological evidence of increasing warfare is widespread, and there is good reason to believe that regional political fragmentation was increasing (Kurjack and Andrews V 1976, Webster 1979). Ball (1979) has suggested that warfare was a response to outside pressure (i.e. the expansion of Chichén Itzá). While this may be true in part, I believe, as do several colleagues, that the situation began to deteriorate long before the expansion of Chichén Itzá. Consequently, there was little to prevent the foreign merchant groups from moving in, which, of course, they did.

Survey data suggest that the people who moved into Yucatán began by deploying themselves along the west and north coasts at locations strategically important for trade and military activities (islands, estuaries, and river entrances). They also appear to have taken over several north coast salt-making localities, long suspected as being one of the prime objectives of these foreign groups. This deployment allowed the intruders, now known as the Itzá, to monopolize the salt trade, which was the most fundamental part of the Yucatecan export economy. The events that followed are well known: the Itzá moved inland, established their capital at Chichén Itzá, and eventually consolidated their control over most of northern Yucatán. They also consolidated their position on the coast, and by monopolizing the salinas and the long-distance trade networks, managed to build at Chichén Itzá one of the largest and wealthiest capitals of Mesoamerica (for a more detailed account of these events, see Andrews 1978b).

The Itzá held sway over most of northern and western Yucatán for approximately 200 years, from A.D. 1000 to 1200. While the bulk of their trading activities were channeled to the southwest, they may have also traded down the Caribbean coast. The uncertainty surrounding the status of the eastern trade networks during the Early Postclassic period is due, in part, to the limited evidence of an Itzá presence along the east coast. From Early Classic times eastern Yucatán was under the control of the city of Cobá, which also appears to have played a major role in the trade networks between the Caribbean and northern Yucatán (Benavides C. 1977, Robles 1977). When Chichén Itzá emerged, Cobá appears to have remained independent; what is not known is whether it retained control of the east coast. The near absence of Itzá remains on the central east coast would suggest that it did; however, future excavations may reveal an Itzá presence at a few ports.

Cobá and Chichén Itzá would appear to have been locked in an economic struggle that Chichén Itzá won. Although failing to conquer Cobá militarily, Chichén Itzá could have strangled it economically by closing down its northern markets and eventually isolating it by taking over the Caribbean trade. In fact, Cobá appears to have faced a serious decline in population, perhaps even a collapse, around A.D. 1100 (Andrews and Robles n.d.).

By this time, it is probable that Itzá traders were moving down the east coast. Two other factors add support to the notion of an Itzá trading presence in the Caribbean. First evidence from northern Belize raises the possibility of an Itzá presence in that area (Chase 1979, Chase and Chase 1982). Second, Chichén Itzá imported huge quantities of jade and metal, a large part of which came from Central America (Proskouriakoff 1974, Bray 1977). The most logical way in which these could have been acquired would have been by means of the east coast trade. Salt and other northern goods must have been offered in exchange.

Shortly after A.D. 1200 a local revolt brought about the fall of Chichén Itzá. At the same time, most of the Itzá trading enclaves of the north and west coasts were also abandoned, a clear indication of a decline of its western trading networks (Andrews 1978b). At this time, too, the eastern trade flourished as never before. Large communities appeared all along the east coast, most of them next to natural harbors that offered excellent facilities for long-distance traders (Andrews IV and Andrews 1975).

The history of the salt trade during the Late Postclassic period is well known from ethnohistoric sources. The city of Mayapán, which succeeded Chichén Itzá as the political center of northern Yucatán, held a monopoly over the northern saltbeds until its own eclipse about A.D. 1450. Thereafter, the salinas were controlled by individual provinces, and the salt trade was managed by members of the elite.

The western trade, though no longer in Itzá hands, continued. When the Spanish arrived, they found that Yucatecan salt was being traded as far as northern Veracruz to the west and the Gulf of Honduras to the southeast.

Southern Sphere

Relatively little is known of the salt trade of the southern sphere during the Postclassic; we do, however, know that several salt sources were bones of contention between competing political groups. The Quiché conquered Sacapulas in the fourteenth century in order to control its salinas, a wise maneuver since the fifteenth-century Cakchiquel revolt probably reduced or cut off their supply of Pacific coastal salt, which I believe the Cakchiquel controlled. Farther west, the Mam kingdom probably incorporated the area around San Mateo Ixtatán and perhaps other Cuchumatan salt-making localities such as San Miguel Acatán and Barillas.

Similar events took place in Chiapas, where, shortly before the conquest, Chiapanecs and Zinacantecos fought over the salinas of Ixtapa.

POST-CONQUEST NATIVE SALT TRADE

Northern Sphere

The native long-distance salt trade networks of the northern sphere gradually faded out after the conquest of Yucatán in 1542, mainly because the Spanish took over the larger salinas and channeled the trade into their own commercial world. Thus, the bulk of exportable Yucatecan salt began to be shipped to central Mexico for use in the silver mines. Some also went to Cuba.

Despite the ladino takeover of the more profitable aspects of the trade, a few remnants of the long-distance native trade persisted into the late sixteenth century and possibly even later. Yucatecan salt continued to be traded to coastal communities along the Gulf coast, beyond Tabasco to Veracruz as far as the Pánuco River (Roys 1943, Stark 1974, Thompson 1966:221).

The eastern trade may have also endured, in limited form, into the early Colonial period; an "underground" trade into Belize and the Petén apparently kept the surviving populations of that area supplied with iron tools and salt until the Spanish conquest of the Petén Itzá in 1697 (G. Jones 1982). We know that the amounts of salt entering the Petén were insufficient, for many groups were resorting to the desperate recourse of burning palms to acquire a salt substitute (López de Cogolludo 1957:700 [Bk. 12, Ch. 7]).

It is difficult to document an underground tropical jungle trade beyond the fringe of colonial authority, but on the basis of what little evidence exists, I believe that the last remnants of the long-distance native salt trade had disappeared by 1700. By that time, the trade appears to have been in ladino hands; also, the British salt trade eventually took over, supplying those areas previously stocked by the east coast trade.

The internal trade of northern Yucatán appears to have survived throughout the Colonial period, but it is difficult to determine whether it was predominantly native or ladino. Until the 1950s, itinerant native merchants supplied many rural communities with salt, and other villages acquired salt by sending workers to the coastal salinas, a pattern that persisted well into the present century. The bulk of the urban trade, however, was handled by ladino producers, merchants, and shopkeepers from early in the Colonial period.

Southern Sphere

The native salt trade networks of the southern sphere were relatively simple logistical operations that endured into the early twentieth century, when they were replaced by the spread of commercial solar salt. Even so, remnants of the trade persist in the highlands. To this day, salt is traded by native merchants from Ixtapa and Atzam in Chiapas, and Sacapulas and San Mateo Ixtatán in Guatemala.

While salt trade and production entail a narrow focus on prehispanic commerce, another major objective of this study was to learn more about Maya trade in general, since the data on salt do hold implications for this broader subject.

From Late Formative times on salt was probably the most important item exchanged in the long-distance Maya trade networks. During the last half of the Late Formative period, however, salt gradually ceased to be a mere trade commodity and became a vital household necessity. The salt trade had become a substantial operation with attendant logistics requiring a high degree of organization. The sheer scope of this enterprise clearly indicates that it could have evolved only within the context of a complex market economy. The Late Formative salt trade undoubtedly involved several thousand tons a year, and to suggest that such amounts were "bartered" would be ludicrous (for a further discussion of the existence of a market economy at this time see Sharer n.d.).

8. Salt, Economics, and Maya Civilization

A civilized life is impossible without salt.
Pliny the Elder, First Century A.D.

EARLY MAYA TRADE INVOLVED THE EXCHANGE of pottery, shell, jade, and obsidian, and possibly salt. This was clearly a "luxury" trade between elites, for none of these goods could be considered essential household items at this early time. Even salt may have been an exotic commodity. Despite the wide trade range of many of these items, the actual volume appears to have been small, a fact which reinforces the notion that it was restricted to limited numbers of people. Whether this early trade was handled through regional exchange networks or by long-distance traders is not known; nor do we have any way of determining whether goods were exchanged through simple barter or a market exchange system. In fact, the emergence of a true market economy in the Maya area is an event that eludes archaeological investigation. The data in this study, however, may shed some light on the problem.

The oldest evidence of long-distance trade in the Maya area dates to Early Formative times (Hammond 1977). By Middle Formative times there is widespread evidence of the existence of far-reaching exchange networks (cf. Adams 1977).

While it appears certain that the volume of the salt trade in Late Formative times precludes the notion of a simple reciprocal system of exchange, we do not know who controlled the mode of production, trade, and distribution of salt. While many scholars may envision a centrally administered system of production and redistribution of household commodities (cf. Pires-Ferreira and Flannery 1976), the logistics of the long-distance salt trade suggest, to me at least, that the trade operated under conditions of market exchange. The early rulers of a large southern community such as Tikal may have administered an internal redistributive economy, but I very much doubt that they had much control over salt production on the north coast of Yucatán, or over the trade mechanisms which brought the salt to Tikal. Thus, the Maya economy of Late Formative times might be viewed as "mixed," in which market-style external trade fed into local redistributive networks. While it may prove difficult to demonstrate the existence of such a system from the available archaeological evidence, the ever-increasing sophistication of analytical models aimed at reconstructing ancient economies offers new avenues of inquiry that may aid significantly in clarifying the nature of prehispanic Maya economic systems (Renfrew 1975, Sharer n.d.).

Not only do the salt data imply the existence of a market economy in Late Formative times, they also point to early emergence of a maritime mercantile trade network. This is not surprising, since seagoing transport and long-distance trading activities date to Olmec times, some ten centuries before the birth of Christ (Velson and Cook 1975, Grove 1968).

In the Maya area the demand for large quantities of salt undoubtedly stimulated the growth of a maritime transport system during the Late Formative period (Freidel 1978). Again, the nature of this enterprise called for specialized long-distance traders who could handle the logistics, a fact which hints at the growth of a powerful merchant class.

Once the long-distance maritime salt trade evolved, it provided a vehicle for the exchange of many other goods which could then be transported in much larger quantities than before. The notion that the salt trade provided the foundations of Maya maritime trade has parallels in several parts of the world (Bloch 1963).

All these processes—the emergence of a market economy, a maritime mercantile trade, and possibly a merchant class—must have had a tremendous impact on the development of Maya civilization.

TRADE AND THE RISE OF MAYA STATES

Several authors, notably Rathje (1971), Tourtellot and Sabloff (1972), Webb (1973, 1975), C. Jones (1979), and Freidel (1978, 1979), see trade as a prime mover in the rise of Maya civilization, a view that is strongly reinforced by this study.

While the earliest trade would appear to have been an exchange of exotic commodities between elites, as Tourtellot and Sabloff (1972) have suggested, I doubt that such a trade gave rise to social complexity among the Maya. It undoubtedly served as a vehicle for establishing important links between elites throughout the Maya area, but such a system of exchange did not call for the large-scale organization that one associates with complex political systems. The emergence of long-distance mercantile salt trade networks, however, does imply the appearance of a complex organizational framework that would have required sophisticated managerial direction. That such an organization emerged in Late Formative times is not surprising: at that time, large, socially complex communities appeared in various localities in the Maya area, such as Uaxactún, Tikal, Becán, Komchén (Dzibilchaltún), Cerros, and Lamanai.

The need for salt, or rather the need for an organizational apparatus that could provide it, was undoubtedly a powerful ingredient in the emergence of pristine states in the Maya area. In short, as Rathje has argued since 1971, organization was likely the key factor.

While this argument provides support for Rathje's (1971b) core–buffer zone hypothesis, I believe his overall model needs revamping. In the first place, there has been some debate over the "essential household" nature of obsidian and basalt implements. Moreover, the model neglects other important core area resources, like cotton, cacao, and a host of "lesser" items. After all, the core area does not appear to be as limited in resources as Rathje originally implied (Dillon 1975, Puleston 1978).

Another feature of the model which should be corrected is the notion that the core area acquired salt from the highlands, an idea suggested by a number of scholars (Smith and Kidder 1951:5, Rathje 1971, Weaver 1972:158, Fox 1978:272–73). The present data clearly negate such a possibility.

A revised core–buffer zone should acknowledge the impact that the central Mexican state of Teotihuacan and its Guatemalan allies in Kaminaljuyú had on the development of the core area (Tourtellot and Sabloff 1972, Webb 1973). While the notion of "secondary state" formation is debatable, one cannot ignore the advantages that would have accrued to those cities with strong connections to the highland megastate. And, interestingly enough, it is a prominent core area site—Tikal—that exhibits the strongest evidence of such connections. As Chris Jones has argued, Tikal exploited this situation by becoming a middleman in the east-west trade of the base of the peninsula (1979).

A further consideration is the role of the core area as a middleman in the north-south trade networks of the Maya area. By controlling the bulk of highland jade and obsidian being traded to the north, cities such as Tikal would have been able to acquire Yucatecan salt and still derive a fair profit from the cost advantage that those items accrued (for a description of the probable economic mechanics of such a trade see Renfrew 1975, Phillips and Rathje 1977).

A revised core–buffer zone hypothesis would view the core area as a dynamic middleman in the long-distance trade networks that criss-crossed the Maya area, rather than as an economically weak region in the business of exporting "cult complexes" (Rathje 1977).

The Flag Follows Salt

Rathje (1971), Webb (1973, 1975), and Renfrew (1975) have argued that access to important natural resources and control of their long-distance trade was a major factor in state formation processes in the ancient world. This was clearly the case among the prehispanic Maya. Many ancient trade networks had one or two key resources whose exchange was the major stimulus for the growth and spread of trade. In the neolithic eastern Mediterranean and Late Formative central Mexican highlands, the resource was obsidian; in the ancient Middle East it was also obsidian, and to a lesser degree, lapis lazuli and chlorite; in the iron age of Europe, and among the earliest kingdoms of China and West Africa, it was salt; among the ancient Maya it was also salt and to a lesser degree cotton, obsidian, and jade.

Access to salt sources or salt trade networks was critical to the security and growth of states in the Maya area, for salt was often the key to larger spheres of trade. In some cases, the desire to control this resource led to outright warfare and conquest: witness the wars between the Chiapanecs and Zinacantecos, the Quiché conquest of Sacapulas, or the Itzá takeover of northern Yucatán. In all of these cases, not only were the salinas at stake but larger trade networks as well.

This pattern of competition over salt sources and trade was clearly a factor in the rise of Postclassic states, but probably had its origins in earlier times. As Rathje has argued, this competition may have spurred the rise of core-area Classic period states as well. But it is improbable that the buffer zone played a passive role in the course of these developments; after all, the key resource, salt, was in the buffer zone. Thus, the development of state organizations in the buffer zone was probably also spurred, in part, by the growth of the salt trade. The emergence of major communities along the north coast of Yucatán during Late Formative times is a clear indication that this was the case. To use a term coined by Renfrew (1975), those communities were "Early State Modules" (ESMs) which evolved into powerful city-states during the Classic period. The same might be argued for coastal or riverine communities, such as Cerros and Lamanai in Belize, which monitored the trade into the core area. In the same way that these communities handled the eastern trade, others must have participated in the western trade. Future investigations will probably reveal one or more major Late Formative ESMs somewhere along the lower Usumacinta River, the main western artery for trade into the core area.

To conclude, I believe that the present data clearly attest to the importance of salt in Maya trade and to the importance of trade among the Maya. Half a century ago, Miguel Othón de Mendizábal (1929) observed that information about important natural resources and knowledge of trade and markets, coupled with economic planning, constitute the fundamental roots essential to the emergence of a strong civilization. The story of the ancient Maya salt trade is a powerful witness to the truth of his words.

Appendix A
Early Seventeenth-Century Salt Production in Yucatán

The production figures reported in the 1603 and 1605 salt surveys (AGI Mexico 72) are tabulated in Table A.1. The 1603 survey was a preliminary one and included only general figures for the four major producing districts. The 1605 survey is very detailed, covering most of the salinas on the north coast of Yucatán. Nonetheless, the reported production figures are in many instances only rough estimates and tend to be rather conservative. For many salinas the documents state that potential production levels are much higher. Moreover, they also acknowledge that many smaller salinas were not properly recorded, and in a few instances, not recorded at all. Also, I suspect that the full production of many salinas was not reported for obvious fiscal reasons. The 1603 survey omits many of the salinas that were in production, and the 1605 survey fails to include several salinas in the Cehpech and Ah Kin Chel provinces. In short, the figures are *very* conservative.

The 1605 survey figures have been grouped by native provinces, as reconstructed by Roys (1957). My figures, however, differ from his in some cases. This may be the result of differing interpretations of the early seventeenth-century paleography of the survey documents, which is extremely difficult to read in places. I have relied heavily on a transcription supplied to me by Ursula Ewald of the University of Heidelberg, but I also consulted the figures cited in the original document.

The major discrepancy between Roys's figures and mine concerns the production of Northern Ah Canul, which he reports as 31,000 fanegas. I am at a loss to explain the discrepancy. It is worth noting, however, that his figure seems more accurate, for the salinas of Northern Ah Canul have traditionally had a level of production comparable to those of Southern Ah Canul. In fact, the former salinas, which today lie in the state of Yucatán, have been the major producers of recent years. Overall, the figures reported by Roys total 166,700 fanegas, or 19,170 tons, 2,000 tons higher than mine.

Finally, there is the matter of converting fanegas into modern equivalents. Roys (1957:15) notes that a fanega "is variously defined as 1.6 bushels, and as a load of 100 pounds of grain." This statement has led a number of scholars to assume that a fanega weighs 100

Table A.1
Early Seventeenth-Century Salt Production in Yucatán

	Fanegas*	Metric Tons
1603 Survey		
Campeche District	78,600	9,039
Mérida District	9,100	1,047
Yobaín District	5,000	575
Valladolid District	58,000	6,670
Total	150,700	17,331
1605 Survey		
Southern Ah Canul		
Campeche District (southern part of Celestún peninsula)	27,500	3,163
Northern Ah Canul		
Mérida District (Celestún north to Punta Boxcohuo)	12,100	1,391
Chakán		
Salinas de Caucel (Punta Choventún to Chuburná)	26,200	3,013
Cehpech		
Laguna Chuburná to Chicxulub	14,700	1,691
Salinas de Dzemul (Xtampú)	600	69
Ah Kin Chel		
Salinas de Sinanché (San Crisanto)	5,000	575
Salinas de Yobaín (Chavihau)	3,000	345
Salinas de Dzidzantún (Santa Clara and Mina de Oro)	2,600	299
Chikinchel/Chauaca		
Valladolid District (Rio Lagartos to El Cuyo)	50,000	5,750
Ecab		
Isla Mujeres	7,000	850
Total	148,700	17,146

*1 fanega = 115 kg

pounds, which is not necessarily the case. The fanega, like the bushel, is a measure of capacity, not weight. Hence, the weight of a fanega will vary enormously depending on the goods being measured by this standard. A fanega of corn may weigh close to 100 pounds, but salt is much heavier.

An extensive survey of the literature on weights and measures used in the Maya area—from Yucatán to El Salvador—suggests that a fanega of salt commonly weighed about 115 kg (Scholes et al. 1936–38, II:66; Zaráte 1917:30; Gutiérrez y Ulloa 1962:135; Suárez Molina 1977, I:100–1, 377). One author has even suggested a weight of 138 kg (De la Peña 1951, IV:1205). Saltmakers on the northwest coast of Chiapas still use the fanega, reckoning its weight at about 120 kg. I have decided to use the more conservative figure of 115 kg as the weight of a fanega of salt.

To conclude, I believe that the 1603 and 1605 survey figures are minimal estimates of the annual production of Yucatán in the early Colonial period. The only exception is the figure for Isla Mujeres, which I believe is greatly exaggerated. Still, this does not alter the picture greatly. I think we can safely assume that the overall reported figures are 10 percent, possibly even 15 percent, less than actual production. Moreover, if we add the potential production of several minor sources around the peninsula, such as Isla del Cármen, Isla Holbox, and Ambergris Cay, we must ultimately recognize that a higher figure is called for. I believe 20,000 tons a year would be an acceptable minimal figure.

Bibliography

Acuña, Rene

1968 Título de los señores de Sacapulas. *Folklore Americas,* No. 28, No. 1, pp.1–45.

Adams, Richard E. W.

1971 The Ceramics of Altar de Sacrificios. *Papers of the Peabody Museum,* Vol. 63, No. 1, Harvard University, Cambridge, Mass.

1973 Fine Orange Pottery as a Source of Ethnological Information. *Contributions of the University of California Archaeological Research Facility.* No. 18, pp. 1–9. Berkeley.

Adams, Richard E. W. (Ed.)

1977 *The Origins of Maya Civilization.* University of New Mexico Press, Albuquerque.

Almquist, Eric L.

1973 Taste and Politics: An Examination of Salt as a Scarce Resource. Master's Thesis, Department of Anthropology, Boston University, Boston.

Andrews, Anthony P.

1977 Reconocimiento arqueológico de la costa norte del estado de Campeche. *Boletín de la Escuela de Ciencias Antropológicas de la Universidad de Yucatán,* Vol. 4, No. 24, pp. 64–77. Mérida.

1978a Breve addenda al "Reconocimiento arqueológico de la costa norte del estado de Campeche." *Boletín de la Escuela de Ciencias Antropológicas de la Universidad de Yucatán,* Vol. 6, No. 33, pp. 40–43. Mérida.

1978b Puertos costeros del Postclásico Temprano en el norte de Yucatán. *Estudios de Cultura Maya.* Vol. 11, pp.75–93. Universidad Nacional Autónoma de México, Mexico D. F.

1980a Salt-making, Merchants and Markets: The Role of a Critical Resource in the Development of Maya Civilization. Doctoral Dissertation, University of Arizona, Tucson.

1980b The Salt Trade of the Ancient Maya. *Archaeology,* Vol. 33, No. 4, pp. 24–33.

n.d. Las salinas de El Salvador. Bosquejo histórico, etnográfico y arqueológico. *Anales del Museo Nacional "David J. Guzmán."* San Salvador. In press.

Andrews, Anthony P. and F. Robles C.
n.d. Chichén Itzá and Cobá. An Itzá-Maya Standoff in Early Postclassic Yucatán. In *The Lowland Maya Postclassic: Questions and Answers,* edited by A.F. Chase and P.M. Rice. University of Texas Press, Austin.

Andrews, E. Wyllys IV
1968 Dzibilchaltún, A Northern Maya Metropolis. *Archaeology,* Vol. 21, No. 1, pp. 36–37.

1969 The Archaeological Use and Distribution of Mollusca in the Maya Lowlands. *Middle American Research Institute,* Publ. 34. Tulane University, New Orleans.

Andrews, E. Wyllys IV and A. P. Andrews
1975 A Preliminary Study of the Ruins of Xcaret, Quintana Roo, Mexico. *Middle American Research Institute,* Publ. 40. Tulane University, New Orleans.

Andrews, E. Wyllys V
1974 Some Architectural Similarities between Dzibilchaltún and Palenque. In *Primera Mesa Redonda de Palenque.* edited by M. Greene Robertson, Part I, pp. 137–48. The Robert Louis Stevenson School, Pebble Beach, California.

1979 Early Central Mexican Architectural Traits at Dzibilchaltún, Yũcatán. *Actes du XLII Congrès International des Americanistes* (Paris 1976), Vol. 8, pp. 237–49. Paris.

1981 Dzibilchaltún. *Handbook of Middle American Indians,* Supplement Vol 1: pp. 313–41. University of Texas Press, Austin.

Andrews, E. Wyllys V et al.
1981 Komchén: An Early Maya Community in Northwest Yucatán. Paper presented at the XVII Mesa Redonda de la Sociedad Mexicana de Antropología, San Cristóbal de las Casas, Chiapas, June 21–27, 1981.

Anonymous
1925 Las costas salvadoreñas. Ligeros apuntes. *Diario La Nación,* April 22. San Salvador.

1944 Salt Industry in Guatemala. *Foreign Commerce Weekly,* Vol. 16, p. 28.

Apenes, Ola
1944 The Primitive Salt Production of Lake Texcoco. *Ethnos,* Vol. 9, No. 1, pp. 25–40.

Archivo General de Centro America, Guatemala (AGCA)
1551 Título de los Señores de Sacapulas. Al, Leg. 6025, Exp. 53.126.

1553 13 de mayo de 1553. Su Majestad solicita informes detallados de las utilidades que obtienen los indios en la explotación de las salinas situadas en la costa del mar del sur, de la provincia de Guatemala. Leg. 1511, p. 184.

1555 Untitled Tribute List. Provincia de Guatemala. Courtesy of Manuel Sánchez Rubio. A3.16, Leg. 2797, Exp. 40466.

1572 18 de mayo de 1572. Que la audiencia informe sobre el estado y número de la salinas que existen en su jurisdicción y si para el fomento de ellas, convendría dárselas a los españoles, quienes solo usarían para el laboreo de la sal. AI.23, Leg. 4575, p. 330.

1573a Ordena su Majestad que sean repartidas las tierras de la costa del mar del sur de la provincia de Guatemala, con el objecto de fomentar la industria salinera. Al. 23, Leg. 1512, p. 428.

1573b 26 de mayo de 1573. Real cédula se ordena al Presidente Dr. Pedro Villalobos, proceda a divulgar el método empleado en Andalucía, para la extracción y elaboración de sal con el fin de abaratar este producto en Guatemala. Al. 23, Leg. 1512, p. 428.

1591a Reales cédulas 11 & 12, directed to the President of the Audiencia of Guatemala, ordering that the salinas of the Guatemalan Pacific Coast be incorporated as part of the Royal Estate, that Indians have free access to their exploitation, and that a tax be imposed on their production. Leg. 6513, p. 717.

1591b Real cédula ordering that all salinas of the New World be incorporated as part of the Royal Estate, and be taxed according to Royal instructions. Al.23, Leg. 1513, p. 724.

1610 Real cédula nullifying the 1603 salt tax. Al.23, Leg. 1514, p. 177.

1672 10 de febrero de 1572. RP. Que a Juan Roque y a sus hijos, vecinos y naturales del pueblo de Nontale, jurisdicción de la ciudad de San Miguel, no sean compelidos a ir en mandamientos de servicios personales a unas salinas. Al.24, Exp. 10208, Leg. 1564, p. 83.

1689 El Común de Tecapa, jurisdicción de San Miguel, El Salvador, piden real provisión de amparo para que no se les compela a ir a trabajar a las salinas que están muy distantes. A.3.12, Leg. 2775.

1690 Los Indios del pueblo de San Juan Comalapa, sobre que el Cacique Don Miguel les exije pasar a la costa a trabajar en sus salinas. Leg. 6061, Exp. 53.974.

1730 Real provisión de amparo a favor de los pueblos de Taxisco, Chiquimulilla y Guazacapán y declárase que los excesos, lagunas y médanos del mar ". . . eran y pertenecían a el uso público y común y que en ellas, ni por ellas tenía interés la Real Corona. . ." Al, Leg. 6008, Exp. 52.922, p. 1.

1779 Revisión de los mojones y tierras de las salinas de Sacapulas, para ser agregado este informe a los autos sobre litigio que sobre tierras seguían los calpules de Santo Domingo Sacapulas. Al, Exp. 53.126, Leg. 6025.

1786–1790 Documents pertaining to the litigation over rights to the lands of the abandoned town of Santa María Magdalena (now Salinas Magdalena). Al, Exp. 53.258, Leg. 6037.

Archivo General de Indias, Seville (AGI)

1548–1551 Libro de Tasaciones de los naturales de las provincias de Guatemala. Nicaragua, Yucatán y Pueblos de Comayagua. (This document is referred to as the "1549 Tribute List" in Yucatán and "1548 Tribute List" in Guatemala and in El Salvador.) Guatemala, Legajo 128.

1603–1608 Survey of the salinas of Yucatán and related documents. Mexico, Legajo 72.

Arias García, Juan Jesús

1972 El grupo doméstico en una localidad henequenera de Yucatán. Tesis Profesional, Escuela Nacional de Antropología e Historia, México, D. F.

Baer, Phillip and W.R. Merrifield

1971 *Two Studies on the Lacandones of Mexico.* Summer Institute of Linguistics, University of Oklahoma, Norman.

Ball, Joseph W.

1974 A Coordinate Approach to Northern Maya Prehistory: 700–1200 A.D. *American Antiquity,* Vol. 39, No. 1, pp. 85–93.

1977a An Hypothetical Outline of Coastal Maya Prehistory: 300 B.C.–A.D. 1200. In *Social Process in Maya Prehistory,* edited by N. Hammond, pp. 167–196. Academic Press, London.

1977b The Rise of the Northern Maya Chiefdoms: A Socio-processual Analysis. In *The Origins of Maya Civilization,* edited by R.E.W. Adams, pp. 101–32. University of New Mexico Press, Albuquerque.

1978 Archaeological Pottery of the Yucatan-Campeche Coast. *Middle American Research Institute* Pub. 46, pp. 69–146. Tulane University, New Orleans.

1979 The 1977 Central College Symposium on Puuc Archaeology: A Summary View. *In the Puuc: New Perspectives,* edited by L. Mills, pp. 46–51. *Scholarly Studies in the Liberal Arts,* Pub. No. 1. Central College, Pella, Iowa.

Ball, Joseph W. and Jack D. Eaton

1972 Marine Resources and the Prehistoric Lowland Maya: A Comment. *American Anthropologist,* Vol. 74, No. 3, pp. 772–76.

Bastow, J. W.

1897 Comercio, moneda y cambio de los antiguos pueblos de México. *XI Proceedings of the International Congress of Americanists* (Mexico 1895), pp. 47–63. México, D. F.

Batterson, Mark and W. W. Bodie

1972 *Salt, the Mysterious Element.* The Dow Chemical Company.

Baudez, Claude F.

1973 Les Camps de saliniers de la côte meridionale de Honduras: donées archaeologiques et documents historiques. In *L'Homme, hier et aujourd'hui. Recueil d'estudes en hommage à André Leroi-Gourhan,* pp. 507–520. Editions Cujas, Paris.

Benavides Castillo, Antonio
1977 Los Caminos prehispánicos de Cobá. *XV Mesa Redonda de la Sociedad Mexicana de Antropología*, (Guanajuato 1977). Vol. 2, pp. 215–225. México, D. F.

Benedict, Francis G. and M. Steggerda
1936 The Food of the Present-day Maya Indians of Yucatan. *Carnegie Institution of Washington,* Publ. No. 456, pp. 155–88. Washington, D.C.

Bloch, M.R.
1963 The Social Influence of Salt. *Scientific American,* Vol. 209, No. 1, pp. 88–98.

Blom, Frans
1932 Commerce, Trade, and Monetary Units of the Maya. *Middle American Research Institute* Publ. No. 4, pp. 531–56. Tulane University, New Orleans.

1945 La Estrella de David en Chiapas. ". . . la sal de la tierra . . ." *Tribuna Israelí,* July 15, México, D. F.

1959 Historical Notes Relating to the Pre-Colombian Amber Trade from Chiapas. *Mitteillungen aus dem Mus. fur Volkerkunde in Hamburg.* Vol. 25, pp. 24–27. Hamburg.

Bonfil Batalla, Guillermo
1962 Diagnóstico sobre el hambre en Sudzal, Yucatán. *Departamento de Investigaciones Históricas,* No. 11. Instituto Nacional de Antropología e Historia, Mexico, D. F.

Borhegyi, Stephan F. de
1958a Hallazgos arqueológicos en aguas del Lago de Amatitlán. *Antropología e historia de Guatemala,* Vol. 10, No. 1, pp. 4–12, Guatemala City.

1958b Underwater Archaeology in Guatemala. *33rd Proceedings of the International Congress of Americanists.* Vol. 2, pp. 229–40. (San Jose 1958), San Jose.

1960 Amatitlán Field Report, 1960. Summary version published in *Katunob,* Vol. 1, No. 4, pp. 67–70.

1965 Archaeological Synthesis of the Guatemala Highlands. *Handbook of Middle American Indians,* Vol. 2, pp. 3–58. University of Texas Press, Austin.

Borhegyi, Stephan F. de and N.S. Scrimshaw
1957 Evidence for Precolumbian Goiter in Guatemala. *American Antiquity,* Vol. 23, No. 2, Pt. 1, pp. 174–76.

Bovill, E. W.
1968 *The Golden Trade of the Moors.* Revised second edition. Oxford University Press, London and New York.

Brand, Donald
1957–1958 *Coastal Study of Southwest Mexico (Coast of Michoacán).* Parts 1 and 2, 2 Vols., Department of Geography, University of Texas, Austin.

Bray, Warwick
1977 Maya Metalwork and its External Connections. In *Social Processes in Maya Prehistory,* edited by N. Hammond, pp. 365–403. Academic Press, New York and London.

Bridbury, A.R.
1955 *England and the Salt Trade in the Later Middle Ages.* Clarendon Press, Oxford.

Brown, Kenneth L.
1977 The Valley of Guatemala: A Highland Port of Trade. In *Teotihuacán and Kaminaljuyú: A Study in Prehistoric Culture Contact,* edited by W. T. Sanders and J. W. Michaels, pp. 205–395. Pennsylvania State University Press, University Park.

Bruce, Robert D.
1968 *Gramática del Lacandón.* Instituto Nacional de Antropología e Historia, México, D. F.

Burbank, Addison
1939 *Guatemala Profile.* Coward-McCann, New York.

Calderón, Jose Tomas
1927 *Prontuario geográfico, commercial, estadístico y servicios administrativos de El Salvador.* San Salvador.

Cancian, Frank
1965 *Economics and Prestige in a Maya Community. The Religious Cargo System in Zinacantan.* Stanford University Press, Palo Alto.

Cardona Lazo, Antonio
1939 *Monografías departamentales (Usulután, San Miguel, Morazán y la Unión).* Publicaciones del Ministerio de Hacienda. Imprenta Nacional, San Salvador.

Cardós de Méndez, Amalia
1959 El comercio de los mayas antiguos. *Acta Antropológica* 2nd Series, Vol. 2, No. 1. Escuela Nacional de Antropología, México, D. F.

Carmack, Robert M.
1973 *Quichean Civilization. The Ethnohistoric, Ethnographic and Archaeological Sources.* University of California Press, Berkeley.

Carrillo y Ancona, Crecensio
1897 El comercio en Yucatán antes del decubrimiento. *XI Proceedings of the International Congress of Americanists* (Mexico 1895), pp. 203–08.

Castro, Juan Francisco
1905 *Geografía elemental de la república de El Salvador.* Tipografía la Unión, San Salvador.

Chamberlain, Robert S.
1947 The Early Years of San Miguel de la Frontera. *Hispanic American Historical Review,* Vol. 27, pp. 623–46.

Chapman, Anne
1957 Port of Trade Enclaves in Aztec and Maya Civilization. In *Trade and Market in the Early Empires,* edited by K. Polanyi, C.M. Arensberg, and H. W. Pearson, pp. 114–53. The Free Press, New York.

Charlton, Thomas H.
1969 Texcoco Fabric-marked Pottery, Tlateles and Saltmaking. *American Antiquity,* Vol. 34, No. 1, pp. 73–76.
1971 Texcoco Fabric-marked Pottery and Saltmaking: A Further Note. *American Antiquity,* Vol. 36, No. 2, pp. 217–18.

Chase, Diane Z.
1979 The Postclassic Period at Nohmul and Santa Rita, Northern Belize. Paper presented to the 1979 Annual Meeting of the American Anthropological Association, Cincinnati, Ohio.

Chase, Diane Z. and A. F. Chase
1982 Yucatec Influence in Terminal Classic Northern Belize. *American Antiquity,* Vol. 47, No. 3, pp. 596–614.

Chinchilla Aguilar, Ernesto
1961 Historia y tradiciones de la ciudad de Amatitlán. *Biblioteca Guatemalteca de Cultura Popular,* Vol. 47, pp. 59–61. Ministerio de Educación Pública, Guatemala City.

Coe, Michael D.
1961 La Victoria. An Early Site on the Pacific Coast of Guatemala. *Peabody Museum Papers,* Vol. 53. Harvard University, Cambridge, Mass.
1962 Preliminary Report on Archaeological Investigations in Coastal Guanacaste, Costa Rica. *34th Proceedings of the International Congress of Americanists,* pp. 358–65. (Vienna 1960). Verlag Ferdinand Berger, Horn, Vienna.

Coe, Michael D. and R.A. Diehl
1980 *In the Land of the Olmec.* 2 vols. University of Texas Press, Austin.

Coe, Michael D. and Kent V. Flannery
1967 Early Cultures and Human Ecology in South Coastal Guatemala. *Smithsonian Contributions to Anthropology,* Vol. 3. Smithsonian Press, Washington, D.C.

Collier, Albert
1964 The American Mediterranean. *Handbook of Middle American Indians,* Vol. 1, pp. 122–42. University of Texas Press, Austin.

Consorcio Salinero de Guatemala
1942–1945 *Informes* (1941–44). AGCA Nos. 2278–2281. Guatemala City.

Conzemius, Eduard
1932 Ethnological Survey of the Miskito and Sumu Indians of Honduras and Nicaragua. *Bureau of American Ethnology, Bulletin, 106.* Smithsonian Institution, Washington, D.C.

Cortés y Larraz, Pedro
1958 Descripción geográfico—moral de la diócesis de Guatemala. Prólogo de don Adrián Recinos (originally ca. 1770). Two volumes. *Biblioteca "Goathemala" de la Sociedad de Geografía e Historia de Guatemala,* Vol. 20, Guatemala City.

Corzo, Angel M.
1943 *Geografía de Chiapas.* Editorial Protos, México, D. F.

Crespo Morales, Marío
1968 Algunos Títulos del Archivo General del Gobierno de Guatemala. Tesis para licenciatura en historia, Universidad de San Carlos, Guatemala.

Criado de Castillo, Alonso
1605 Autos. . . sobre. . . la reducción de los Indios infieles llamados tequequas comarcanos al dicho Puerto de Amatique. Archivo General de Indias, Guatemala 129, Seville.

Culbert, T. Patrick
1977 Maya Development and Collapse: An Economic Perspective. In *Social Processes in Maya Prehistory,* edited by N. Hammond, pp. 509–30. Academic Press, New York and London.

Culbert, T. Patrick (Ed.)
1973 *The Classic Maya Collapse.* A School of American Research Book, University of New Mexico Press, Albuquerque.

Dampier, William
1906 *Dampier's Voyages.* E. Grant Richards, London.

Dauphinee, James A.
1960 Sodium Chloride in Physiology, Nutrition and Medicine. In *Sodium Chloride: The Production and Properties of Salt and Brine,* edited by Dale W. Kaufman, pp. 382–453. Hafner Publishing Co., New York.

Deevey, Edward S.
1957 Limnologic Studies in Middle America. *Connecticut Academy of Arts and Sciences,* Vol. 39, pp. 213–328. Yale University Press, New Haven.

Deevey, E.S., D.S. Rice, P.M. Rice and Others
1979 Mayan Urbanism: Impact on a Tropical Karst Environment. *Science*, Vol. 206, pp. 298–306.

De La Peña, Moisés T.
1951 *Chiapas económico,* Vol. 4, pp. 1203–11. Departamento de Prensa y Turismo, Sección Autográfica, Tuxtla Gutiérrez, Chiapas.

Díaz del Castillo, Bernal
1967 *Historia verdadera de las cosas de la Nueva España.* Editorial Porrua, México, D. F.

Dieseldorff, Erwin P.
1909 Klassifizierung einer archäologischen funde in nordlichen Guatemala. *Zeitschrift fur Ethnologie,* Vol. 41, pp. 862–74. Berliner Gesselschoft fur Anthropologie, Ethnologie und Urgeschichte, Berlin.

Dillon, Brian
1975 Notes on Trade in Ancient Mesoamerica. *Contributions of the University of California Archaeological Facility,* No. 24, pp. 80–135. Berkeley.
1977 Salinas de los Nueve Cerros, Alta Verapaz, Guatemala. Preliminary Archaeological Investigations. *Ballena Press Studies in Mesoamerican Art, Archaeology and Ethnohistory,* No. 2. Socorro, New Mexico.
1979 The Archaeological Ceramics of Salinas de los Nueve Cerros, Alta Verapaz, Guatemala. Doctoral dissertation, University of California, Berkeley.

Dirección General de Estadísticas y Censos (DGEYC)
1921–1976 *Anuarios estadísticos.* San Salvador.

Dollfus, A. and Mont-Serrat, E. De
1868 *Voyage géologique dans les républiques de Guatemala et de Salvador.* Paris.

Drennan, Robert D.
1976 Fábrica San José and Middle Formative Society in the Valley of Oaxaca. In *Prehistory and Human Ecology of the Valley of Oaxaca,* edited by Kent Flannery, Vol. 4. Memoirs of the Museum of Anthropology, No. 8. University of Michigan, Ann Arbor.

Drucker, Philip
1948 Preliminary Notes on an Archaeological Survey of the Chiapas Coast. *Middle American Research Institute,* Publ. No. 15, pp. 151–64. Tulane University, New Orleans.

Eaton, Jack D.
1974 Shell Celts from Coastal Yucatán. *Texas Archaeological Society,* Bulletin 45, pp. 197–208.
1978 Archaeological Survey of the Yucatan-Campeche Coast. *Middle American Research Institute,* Publ. 46, pp. 1–67, Tulane University, New Orleans.

Echánove, Policarpio Antonio and Pedro Bolio
1849 Cuadro estadístico de Yucatán en 1814. *El Fénix,* Nos. 21–25; 27–29 (10 de febrero al 20 de marzo). Campeche.

Edwards, Clinton
1954 Geographical Reconnaissance in the Yucatan Peninsula. Department of Geography, University of California, Berkeley.

Eskew, G. L.
1948 *Salt. The Fifth Element.* J. G. Ferguson, Chicago.

Estrada, Juan de and Fernando de Niebla
1955 Descripción de la provincia de Zapotitlán y Suchitepequez. Año de 1579. *Anales de la Sociedad de Geografía e Historia,* Vol. 28, Nos. 1–4, pp. 68–83. Guatemala City.

Feldman, Lawrence H.
1971 A Tumpline Economy: Production and Distribution Systems of Early Central-East Guatemala, Doctoral Dissertation, Pennsylvania State University, University Park.

Feldman, Lawrence H. (*continued*)

1972 An Old Título of Santo Domingo Sacapulas, Guatemala. *Katunob,* Vol. 8, No. 1, pp. 70–77.

1973 Chiapas in 1774 (transcription of a *Relación* of the Obispado de Chiapas, written in 1774). *Contributions of the University of California Archaeological Research Facility,* No. 18, pp. 105–35. Berkeley.

1975 Riverine Maya. The Torquegua and Other Chols of the Lower Motagua Valley. *Museum Brief* No. 15, University of Missouri, Columbia.

1978 Moving Merchandise in Protohistoric Central Quauhtemallan. In *Mesoamerican Communication Routes and Cultural Contacts,* edited by Thomas A. Lee, Jr. and Carlos Navarrete, pp. 7–17. *Papers of the New World Archaeological Foundation* No. 40, Brigham Young University, Provo, Utah.

Fox, John W.

1978 *Quiché Conquest. Centralism and Regionalism in Highland Guatemalan State Development.* University of New Mexico Press, Albuquerque.

Freidel, David

1976 Late Postclassic Settlement Patterns on Cozumel Island, Quintana Roo, Mexico. Doctoral Dissertation, Harvard University, Cambridge, Mass.

1978 Maritime Adaptation and the Rise of Maya Civilization. In *Prehistoric Coastal Adaptations. The Economy of Maritime Middle America,* edited by B. L. Stark and B. Voorhies, pp. 239–65. Academic Press, New York.

1979 Culture Areas and Interaction Spheres: Contrasting Approaches to the Emergence of Civilization in the Maya Lowlands. *American Antiquity,* Vol. 44, No. 1, pp. 36–54.

Fuentes y Guzmán, Francisco Antonio

1932–1933 *Recordación Florida. Discurso historial y demostración natural, mineral, militar y política del Reyno de Guatemala.* Biblioteca "Goathemala," Vols. 6–8, Sociedad de Geografía e Historia, Guatemala City.

Gage, Thomas

1928 *The English-American. A New Survey of the West Indies, 1648,* edited with an Introduction by A. P. Newton. The Broadway Travelers, George Routledge & Sons, London.

Gall, Francis (Ed.)

1968 El Licensiado Francisco Marroquín y una descripción de El Salvador, año de 1532. *Anales de la Sociedad de Geografía e Historia,* Vol. 41, Nos. 2–4, pp. 199–232. Guatemala City.

Gann, Thomas W. F.

1918 The Maya Indians of Southern Yucatan and Northern British Honduras. Smithsonian Institution, *Bulletin of the Bureau of American Ethnology* No. 64, Washington, D.C.

Gann, T. W. F. and J. E. S. Thompson

1931 *The History of Maya.* Charles Scribner's Sons, London.

García de Palacio, Lic. Diego
1927 Relación hecha por el Licensiado Palacio al Rey D. Felipe II en la que describe la provincia de Guatemala, las costumbres de los indios y otras cosas notables (Orig. 1576). *Anales de la Sociedad de Geografía e Historia,* Vol. 4, No. 1, pp. 71–92. Guatemala City.

Garza T. de González, Silvia and E.B. Kurjack
1980 *Atlas arqueológico del estado de Yucatán.* Instituto Nacional de Antropología e Historia, México, D. F.

Gibson, Charles
1964 *The Aztecs under Spanish Rule.* Stanford University Press, Palo Alto.

Gilmore, Harlan W.
1955 Cultural Diffusion via Salt. *American Anthropologist,* Vol. 57, No. 5, pp. 1011–15.

Godelier, Maurice
1971 "Salt Currency" and the Circulation of Commodities Among the Baruya of New Guinea. In *Studies in Economic Anthropology,* edited by G. Dalton, pp. 52–73. *Anthropological Studies* No. 7, American Anthropological Association, Washington, D.C.

Goncalves de Lima, Oswaldo
1955 El Empleo de la sal de cenizas por los indígenas brasileños. *El México Antiguo,* Vol. 8, pp. 353–64. México, D. F.

González de Cossío, F.
1952 *El Libro de las tasaciones de pueblos de la Nueva España, siglo XVI.* Archivo General de La Nación, México, D. F.

Goubaud Carrera, Antonio, J. D. González and S. Tax
1944 Reconnaissance of Northern Guatemala, 1944. *Microfilm Collection of Manuscripts on Middle American Cultural Anthropology,* No. 17, University of Chicago.

Grollig, Francis X, S.J.
1962 The Vanishing Folk-religion of the Highlands of Guatemala. *34th Proceedings of the International Congress of Americanists,* pp. 183–86 (Vienna 1960), Vienna.

Grollman, Arthur
1961 The Role of Salt in Health and Disease. *American Journal of Cardiology,* Vol. 8. pp. 593–602.

Grove, David C.
1968 The Pre-Classic Olmec in Central Mexico: Site Distribution and Inferences. In *Dumbarton Oaks Conference on the Olmec,* edited by E. P. Benson, pp. 179–85. Dumbarton Oaks, Washington, D.C.

Gutiérrez, H., Jesús Agripino
1949 *Ixtapa. Estampas de mi pueblo.* Tuxtla Gutiérrez, Chiapas.

Gutiérrez y Ulloa, Antonio
1962 Estado general de la provincia de San Salvador: Reyno de Guatemala (año de 1807). *Colección Historia,* Vol. 7. Ministerio de Educación, Dirección General de Publicaciones, San Salvador.

Gutman, Theodore
1972 Review of the Importance of Salt in Historical Literature with Special Reference to West Mexico. *Katunob,* Vol. 8, No. 6, pp. 40–49.

Habel, S.
1878 The Sculptures of Santa Lucia Cosumalwhuapa in Guatemala. *Smithsonian Contributions to Knowledge,* Vol. 22, No. 269. Smithsonian Institution, Washington, D.C.

Hammond, Norman
1976 Maya Obsidian Trade in Southern Belize. In *Maya Lithic Studies.* Belize Field Symposium, edited by T. R. Hester and N. Hammond, pp. 71–81. *Center for Archaeological Research,* Special Report No. 4. University of Texas at San Antonio.
1977 The Earliest Maya. *Scientific American,* Vol. 236, No. 3, pp. 116–33.

Harris, Florence E.
1960 Economics of the Salt Industry. In *Sodium Chloride: The Production and Properties of Salt and Brine,* edited by Dale W. Kaufmann, pp. 627–61. Hafner Publishing Co., New York.

Haviland, William A.
1965 Prehistoric Settlement at Tikal, Guatemala. *Expedition,* Vol. 7, No. 3, pp. 15–23.
1967 Stature at Tikal, Guatemala: Implications for Ancient Maya Demography and Social Organization. *American Antiquity,* Vol. 32, No. 3, pp. 316–25.
1970 Tikal, Guatemala and Mesoamerican Urbanism. *World Archaeology,* Vol. 2, pp. 186–98.
1975 The Ancient Maya and the Evolution of Urban Society. *Katunob,* Miscellaneous Series, No. 7.

Heizer, Robert F. and Henry Rapoport
1962 Identification of Distichlis Salt Used by California Indians. *Masterkey,* Vol. 36, No. 4, pp. 146–48.

Helbig, Karl M.
1964a *La Cuenca superior del Río Grijalva. Un Estudio regional de Chiapas, Sureste de México.* Instituto de Ciencias y Artes de Chiapas, Tuxtla Gutiérrez, Chiapas.
1964b *El Soconusco y su zona cafetalera en Chiapas.* Instituto de Ciencias y Artes de Chiapas, Tuxtla Gutiérrez, Chiapas.

Helfritz, Hans
1963 Vorkolumbische Salzgewinnung in Guatemala. *Kosmos,* 59 Jahrgang, Heft 7, pp. 300–05. Stuttgart.

Hester, J. A.
1953 Agriculture, Economy and Population Densities of the Maya. *Carnegie Institution of Washington Yearbook,* Vol. 52, pp. 288–92.

Hidalgo, Joseph Domingo
1952 Memoria para hacer una descripción puntual del Reino de Guatemala. (Orig. 1797–98). *Anales de la Sociedad de Geografía e Historia,* Vol. 26, Nos. 3–4, pp. 383–413. Guatemala City.

Holmberg, Allen P.
1969 *Nomads of the Long Bow: The Siriono of Eastern Bolivia.* Natural History Press, Garden City.

Hooton, Earnest A.
1940 Skeletons from the Cenote of Sacrifice at Chichen Itza. In *The Maya and Their Neighbors,* pp. 272–80. D. Appleton-Century, New York and London.

Hunt, Eva
1972 Irrigation and the Socio-political Organization of Cuicatec Cacicazgos. In *The Prehistory of the Tehuacan Valley,* Vol. 4, *Chronology and Irrigation,* edited by R. S. MacNeish, pp. 162–259. University of Texas Press, Austin.

Hunter, H. V.
1940 The Ethnography of Salt in Aboriginal North America. Master's Thesis, University of Pennsylvania, Philadelphia.

Isphording. W. C.
1975 The Physical Geology of Yucatan. *Transactions, Gulf Coast Association of Geological Societies,* Vol. 25, pp. 231–62.

Jones, Christopher
1979 Tikal as a Trading Center: Why it Rose and Fell. Paper presented to the 43rd Proceedings of the International Congress of Americanists, August 10–17, Vancouver, B.C.

Jones, Grant D.
1982 Agriculture and Trade in the Colonial Period Central Maya Lowlands. In *Maya Subsistence,* edited by K.V. Flannery, pp. 275–93. Academic Press, New York.

Juarros, Domingo
1823 *A Statistical and Commercial History of the Kingdom of Guatemala in Spanish America,* translated by J. Baily. London.

1936 *Compendio de la historia de la ciudad de Guatemala* (orig. 1809). Tipografía Nacional, Guatemala City.

Judson, Lyman and Ellen Judson
1949 *Let's Go to Guatemala.* Harper and Bros., New York.

Kaufmann, Dale W. (Ed.)
1960 *Sodium Chloride: The Production and Properties of Salt and Brine.* Hafner Publishing Co., New York.

Kaunitz, H.
1956 Causes and Consequences of Salt Consumption. *Nature,* Vol. 178, pp. 1141–44.

Kidder, Alfred V.
1940 Pottery from Champerico, Guatemala. *Carnegie Institution of Washington,* Vol. 1, No, 2, pp. 4–10. Washington, D.C.

Kurjack, Edward B. and E. W. Andrews V
1976 Early Boundary Maintenance in Northwest Yucatan, Mexico. *American Antiquity,* Vol. 41, No. 3, pp. 318–25.

LeFarge, Oliver II and Douglas Byers
1931 The Year Bearer's People. *Middle American Research Institute* Publ. 3. Tulane University, New Orleans.

Lange, Frederick W.
1971 Marine Resources: A Viable Subsistence Alternative for the Prehistoric Lowland Maya. *American Anthropologist,* Vol. 73, No. 3, pp. 619–39.

Lefond, Stanley J.
1969 *Handbook of World Salt Resources.* Plenum Press, New York.

LePlongeon, Alice
1889 *Here and There in Yucatan.* John W. Lovell Co., New York.

Litke, Gene R.
1975 The Stratigraphy and Sedimentation of the Barillas Quadrangle, Department of Huehuetenango, Guatemala, C. A. Master's Thesis, University of Texas at Arlington.

Lonn, Ella
1965 Salt as a Factor in the Confederacy. *Southern Historical Publication* No. 4, University of Alabama Press, Tuscaloosa.

López, Lorenzo
1974 *Estadística general de la República de El Salvador.* Dirección de Publicaciones, Ministerio de Educación, San Salvador.

López de Cogolludo, Fray Diego
1957 *Historia de Yucatán* (5th Edition; edited with notes by J. I. Rubio Mañe), 2 vols. Editorial Academia Literaria, México, D. F.

Lowe, Gareth W.
1959 Archaeological Explorations of the Upper Grijalva River, Chiapas, Mexico. *Papers of the New World Archaeological Foundation* No. 2. Brigham Young University, Provo, Utah.

1975 The Early Preclassic Barra Phase of Altamira, Chiapas. A Review with New Data. *Papers of the New World Archaeological Foundation* No. 38. Brigham Young University, Provo, Utah.

Lozano García, Raúl
1946 *Estudio tecnológico de la industria de la sal en México* Instituto de Geología, Universidad Nacional Autónoma de México, México, D. F.

McBryde, Felix W.
1933 Sololá: A Guatemalan Town and Cakchiquel Market Center. *Middle American Research Institute* Pub. 5, pp. 45–152. Tulane University, New Orleans.

1947 Cultural and Historical Geography of Southwest Guatemala. *Institute of Social Anthropology* Publ. 4, Smithsonian Institution, Washington, D.C.

McCullough, John M.
1973 Human Ecology, Heat Adaptation, and Belief Systems: The Hot-Cold Syndrome of Yucatan. *Journal of Anthropological Research,* Vol. 29, No. 1, pp. 32–36.

MacLeod, Murdo J.
1973 *Spanish Central America. A Socioeconomic History, 1520–1720.* University of California Press, Berkeley.

MacNeish, R. S., F. A. Peterson and K. V. Flannery
1970 Ceramics. In *The Prehistory of the Tehuacan Valley,* Vol. 3, edited by R. S. MacNeish. University of Texas Press, Austin.

MacNeish, R. S., F. A. Peterson and J. A. Neely
1972 The Archaeological Reconnaissance. In *The Prehistory of the Tehuacán Valley,* Vol. 5, edited by R.S. MacNeish, pp. 341–495. University of Texas Press, Austin.

McVicker, Donald E.
1969 The Place of the Salt: Archaeological Survey and Excavations in the Valley of Ixtapa, Chiapas, Mexico. Doctoral Dissertation, University of Chicago, Chicago.

1974 Variation in Protohistoric Maya Settlement Pattern. *American Antiquity,* Vol. 34, No. 4, pp. 546–56.

1978 Prehispanic Trade in Central Chiapas, Mexico. In *Mesoamerican Communication Routes and Cultural Contacts,* edited by Thomas A. Lee, Jr. and Carlos Navarrete, pp. 176–86. *Papers of the New World Archaeological Foundation* No. 40. Brigham Young University, Provo, Utah.

n.d. A Preliminary Archaeological Survey of the Municipio of Zinacantan, Chiapas, Mexico (1972). MS on file, Harvard University Chiapas Project, Cambridge, Mass.

Manó, J. Carlos
1940 Tercer informe presentado a la Secretaría de Fomento: Salinas de Magdelena (1883). *Anales de la Sociedad de Geografía e Historia de Guatemala,* Vol. 16, No. 6, pp. 502–11. Guatemala City.

Martínez Vera, Rogerio
1979 Los Impuestos procedentes de la época colonial en el ramo de la sal. *Novedades de Yucatán,* April 15, Mérida.

Mata Amado, G.
1975 *Arqueología sub-acuática, Amatitlán, Guatemala.* Asociación Tikal, Guatemala City.

Mayer-Oakes, William J.
1975 A Stratigraphic Excavation at El Risco, Mexico. *Proceedings of the American Philosophical Society,* Vol. 103, No. 3, pp. 332–33, 365–72. Philadelphia.

Medel y Alvarado, León
1963 *Historia de San Andrés Tuxtla.* 2 vols. Colección Suma Veracruzana. Serie Historiografía. Editorial Citlapetl, México, D. F.

Mendizábal, Miguel Othón de
1929 *Influencia de la sal en la distribución geográfica de los grupos indígenas de México.* Imprenta del Museo Nacional de Arqueología, Historia y Etnografía, México, D. F.

1930 Influencia de la sal en la distribución geográfica de los grupos indígenas de México. *Proceedings 23rd International Congress of Americanists,* pp. 93–100. New York.

Mendizábal, Miguel Othón de (*continued*)

1946a Influencia de la sal en la distribución geográfica de los grupos indígenas de México. *Obras Completas,* Vol. 2, pp. 181–340, México, D. F.

1946b La distribución geográfica de la sal. *Mexico Prehispánico,* pp. 742–53. México, D. F.

Molina Solís, Juan Francisco

1943 *Historia del descubrimiento y conquista de Yucatán.* Ediciones Mensaje, México, D. F.

Moriarty, James R.

1964–1965 The Influence of Strand Plain Morphology on the Development of Primitive Industries along the Costa de Nayarit, Mexico. *America Indigena,* Vol. 24, No. 4, pp. 365–79, Vol. 25, No. 1, pp. 65–77.

Mullerried, F. K. G.

1957 *La geología de Chiapas.* Gobierno Institucional del Estado de Chiapas, Tuxtla Gutiérrez, Chiapas.

Multhauf, Robert P.

1978 *Neptune's Gift: A History of Common Salt.* The Johns Hopkins University Press, Baltimore.

Municipio de Cobán, Alta Verapaz

1882–1911 Titulo librado a favor de la municipalidad de Cobán del terreno nombrado Nueve Cerros compuesto de 1530 hectaras, 94 áreas y noventa metros cuadrados y de concesión de las salinas que en el existen. Archives of the Municipality of Cobán, Alta Verapaz.

Navarrete, Carlos

1959 A Brief Reconnaissance in the Region of Tonalá, Chiapas, Mexico. *Papers of the New World Archaeological Foundation* No. 4, Brigham Young University, Provo, Utah.

1966 The Chiapanec History and Culture. *Papers of the New World Archaeological Foundation* No. 21. Brigham Young University, Provo, Utah.

1974 The Olmec Rock Carvings at Pijijiapan, Chiapas, Mexico and other Olmec Pieces from Chiapas and Guatemala. *Papers of the New World Archaeological Foundation* No. 35. Brigham Young University, Provo, Utah.

Navarrete, Carlos and Thomas A. Lee, Jr.

1969 Apuntes sobre el trabajo del ambar en Simojovel, Chiapas. *Instituto Nacional de Antropología e Historia* Boletín 35, pp. 13–19. México, D. F.

Nelson, Fred W.

1980 Rutas de intercambio en la península de Yucatán en las diferentes épocas arqueológicas según la evidencia de la obsidiana. *XVI Mesa Redonda de la Sociedad Mexicana de Antropología* (Saltillo 1979), pp. 349–55. México, D. F.

Nenquin, J.

1961 Salt: A Study in Ecomonic Prehistory. Dissertationes Archaeologique. Gandenses Brugge, De Tempel.

Noguera, Eduardo

1975 Identificación de una saladera. *Anales de Antropología,* Vol. 12, pp. 117–51. Universidad Nacional Autónoma de México, México, D. F.

Noyes, Ernest

1932 Fray Alonso Ponce in Yucatán, 1588. Report written by Fray Antonio de Ciudad Real. *Middle American Research Series,* Publ. 4, pp. 297–372. Middle American Research Institute, Tulane University, New Orleans.

Oakes, Maud

1951 *The Two Crosses of Todos Santos.* Pantheon Books, New York.

Ober, Frederick A.

1887 *Travels in Mexico and Life Among the Mexicans.* Estes and Lauriat, Boston.

Olavarreta, Joseph de

1935 Relación geográfica del Partido de Huehuetenango—1740. *Boletín del Archivo General del Gobierno,* Vol. 1, No. 1, pp. 16–24. Guatemala City.

Paillés H., Maricruz

1978 The Process of Transformation at Pajón: A Preclassic Society Located in an Estuary in Chiapas, Mexico. In *Prehistoric Coastal Adaptations. The Economy of Maritime Middle America,* edited by B. L. Stark and B. Voorhies, pp. 81–95. Academic Press, New York.

Pardo, J. Joaquín (Ed.)

1937 Año de 1763—Autos formados sobre la Real Cédula para que esta Real Audiencia con la brevedad y recerva posible remita una relación individual de los Corregimientos y Alcaldías Mayores de este Reyno. *Boletín del Archivo General del Gobierno,* Vol. 2, No. 3, pp. 274–329 y No. 4, pp. 448–86. Guatemala City.

Parsons, Jeffrey R.

1971 Prehistoric Settlement Patterns in the Texcoco Region, Mexico. *Memoirs of the Museum of Anthropology,* No. 3. University of Michigan, Ann Arbor.

Paso y Troncoso, F. Del (Ed.)

1905–1906 *Papeles de la Nueva España.* Madrid. (6 vols.)

1939–1942 *Epistolario de Nueva España, 1505–1818.* Biblioteca Historia Mexicana de Obras Inéditas, 2nda. Serie, Mexico. (16 vols.)

Peterson, David A.

1976 Ancient Commerce. Doctoral Dissertation, State University of New York, Binghamton.

1977 El fin de la época clásica: cambios prehistóricos en el valle de Oaxaca y sus relaciones con cambios prehistóricos en el valle de Teotihuacán. *XV Mesa Redonda de la Sociedad Mexicana de Antropología,* Vol. 2, pp. 369–75. Guanajuato, Mexico.

Pettersen, Carmen L.

1976 *Maya of/de Guatemala. Life and Dress. Vida y Traje.* Bilingual English-Spanish edition. Guatemala City.

Phillips, David A.

1979 Material Culture and Trade of the Postclassic Maya. Doctoral Dissertation, University of Arizona, Tucson.

Phillips, David A. and William L. Rathje

1977 Streets Ahead: Exchange Values and the Rise of Maya. In *Social Processes in Maya Prehistory,* edited by N. Hammond, pp. 103–12. Academic Press, New York and London.

Piña Chan, Román

1978 Commerce in the Yucatán Peninsula: The Conquest and Colonial Period. In *Mesoamerican Communication Routes and Cultural Contacts,* edited by Thomas A. Lee, Jr. and Carlos Navarrete, pp. 37–48. *Papers of the New World Archaeological Foundation* No. 40. Brigham Young University, Provo, Utah.

Pineda, Emeterio

1845 *Descripción geográfica del departamento de Chiapas y Soconusco.* Imprenta de Ignacio Cumplido, México, D. F.

Pires-Ferreira, Jane W.

1975 Formative Mesoamerican Exchange Networks with Special Reference to the Valley of Oaxaca. In *Prehistory and Human Ecology of the Valley of Oaxaca,* edited by K. V. Flannery, Vol. 3. Memoirs of the Museum of Anthropology No. 8. University of Michigan, Ann Arbor.

Pires-Ferreira, Jane W. and K. V. Flannery

1976 Ethnographic Models for Formative Exchange. In *The Early Mesoamerican Village,* edited by K. V. Flannery, pp. 286–292. Academic Press, New York.

Pohl, Mary E. D.

1976 Ethnozoology of the Maya: An Analysis of Fauna from Five Sites in the Peten. Doctoral Dissertation, Harvard University, Cambridge, Mass.

Pollock, H. E. D. and C. E. Ray

1957 Notes on Vertebrate Animal Remains from Mayapan. *Carnegie Institution of Washington,* Current Reports No. 41. Washington, D.C.

Ponce, Fray Alonso

1873 Relación breve y verdadera de algunas cosas de las muchas que sucedieron al Padre Fray Alonso Ponce en las provincias de la Nueva España, escrita por dos religiosos, sus compañeros. (1588, 2 vols.). *Colección de Documentos Inéditos para la Historia de España,* Vols. 57–58. Imprenta de la Viuda de Calero, Madrid.

Pozas, Ricardo

1959 Chamula: un pueblo indio de los Altos de Chiapas. *Memorias, Instituto Nacional Indigenista,* Vol. 18. México, D. F.

1977 Chamula: un pueblo indio de los Altos de Chiapas. *Instituto Nacional Indigenista.* México, D. F., (2nd edition).

Proskouriakoff, Tatiana

1974 Jades from the Cenote of Sacrifice, Chichen Itza, Yucatan. *Memoirs of the Peabody Museum,* Vol. 10, No. 1. Harvard University, Cambridge, Mass.

Puleston, Dennis

1978 Terracing, Raised Fields and Tree Cropping in the Maya Lowlands: A New Perspective on the Geography of Power. In *Prehispanic Maya Agriculture,* edited by P.D. Harrison and B. L. Turner, pp. 225–45.

Quiñonez, José A.

1930 *Directorio general de la República de Guatemala.* Tipografía Nacional, Guatemala City.

Quirós, José María

1973 Memoria de estatuto (1817). In *Descripciones Económicas Regionales de Nueva España,* edited by E. Florescano and I. Gil Sánchez, pp. 231–64. Departamento de Investigaciones Históricas, Instituto de Antropología e Historia, México, D. F.

Rathje, William L.

1971 The Origin and Development of Lowland Classic Maya Civilization. *American Antiquity,* Vol. 36, No. 3, pp. 275–85.

1972 Praise the Gods and Pass the Metates: A Hypothesis of the Development of Lowland Rainforest Civilizations. In *Contemporary Archaeology. A Guide to Theory and Contributions,* edited by M. P. Leone, pp. 365–92. Southern Illinois University Press, Carbondale and Edwardsville.

1977 The Tikal Connection. In *The Origins of Maya Civilization,* edited by R. E. W. Adams, pp. 373–82. University of New Mexico Press, Albuquerque.

Recinos, Adrián

1913 *Monografía del Departamento de Huehuetenango, República de Guatemala.* Tipografía Sánchez y De Guisse, Guatemala City.

1954 Monografía del departamento de Huehuetenango (2nd edition). *Colección Monográfica,* No. 2. Editorial del Ministerio de Educación Pública, Guatemala City.

Redfield, Robert and Margaret P. Redfield

1940 Disease and its Treatment in Dzitás, Yucatán. *Carnegie Institution of Washington* Publ. 523, Contribution 32, pp. 49–81. Washington, D.C.

Redfield, Robert and A. Villa Rojas

1934 Chan Kom. A Maya Village. *Carnegie Institution of Washington,* Publ. 448. Washington, D.C.

Regil, Pedro Manuel

1848–1849 Memoria instructiva sobre al comercio general de la provincia de Yucatán, y particular del puerto de Campeche, formada por el Sr. Don Pedro Manuel Regil. (Escrita en 1811). *El Fénix Nos.* 7 – 10, 14 – 17; 21 (1 de diciembre de 1848 al 10 de febrero de 1849). Campeche.

Regil, José M. and Alonso M. Peón

1853 Estadística de Yucatán. *Boletín de la Sociedad Mexicana de Geografía y Estadística,* Epoca 1, Vol. 3 (1852), pp. 237–340. México, D. F.

Reina, Ruben E. and R. M. Hill, II

1978 *The Traditional Pottery of Guatemala.* University of Texas Press, Austin.

Reina, Ruben E, and John Monaghan
1981 The Ways of the Maya. Salt Production in Sacapulas, Guatemala. *Expedition,* Vol. 23, No. 3, pp. 13–33.

Relaciones de Yucatán (RY)
1898–1900 In *Colección de documentos inéditos relativos al descubrimiento, conquista y organización de las antiguas posesiones españolas de ultramar.* 2nd Series, Vols. 2 (1898; *RY I*) and 13 (1900; *RY II*). Madrid.

Remesal, A. de
1932 *Historia general de las Indias occidentales y particular de la gobernación de Chiapas y Guatemala* (Orig. 1619). 2 Vols. Guatemala City.

Renfrew, Colin
1975 Trade as Action at a Distance: Questions of Integration and Communication. In *Ancient Civilizations and Trade,* edited by J. A. Sabloff and C. C. Lamberg-Karlovsky, pp. 3–59. University of New Mexico Press, Albuquerque.

Républica de Guatemala, Dirección General de Estadística (RG, DGE)
1930–1943 *Memorias.* Tipografía Nacional, Guatemala City.
n.d. Salt Production Statistics, Pacific Coast Industry, 1950–1972. Guatemala City.

República de Guatemala, Ministerio de Fomento (RG, MF)
1881–1921 *Memorias.* Guatemala City.

República de Guatemala, Ministerio de Hacienda y Crédito Público (RG, MH Y CP)
1880–1955 *Memorias.* Guatemala City.

República de Guatemala, Secretaría de Fomento, Sección de Estadística (RG, SF)
1884 *Anales estadísticos de la república de Guatemala, año de 1883.* Tipografía "El Progreso," Guatemala City.

Riehm, Karl
1961 Prehistoric Salt-boiling. *Antiquity,* Vol. 35, No. 139, pp. 181–91.

Roberts, D. F. and M. T. Newman
1970 Physiological Studies. *Handbook of Middle American Indians,* Vol. 9, pp. 148–63. University of Texas Press, Austin.

Robles Castellanos, Fernando
1977 Evidence for Late Classic Political Structures Among the Maya: The Sacbes of Northern Yucatan. Paper presented to the 42nd Annual Meeting of the Society for American Archaeology, April 28–30, New Orleans.

Rovner, Irwin
1975 The Cyclical Rise and Fall of Maya Lithic Trade Spheres. Paper presented to the 40th Annual Meeting of the Society for American Archaeology, May 8–10. Montreal, Canada.

Roys, Ralph L.
1931 The Ethno-Botany of the Maya. *Middle American Research Series,* Publ. 2, Tulane University, New Orleans.
1933 The Book of Chilam Balam of Chumayel. *Carnegie Institution of Washington,* Publ. 505, Washington, D.C.
1943 The Indian Background of Colonial Yucatan. *Carnegie Institution of Washington,* Publ. 548. Washington, D.C.
1957 The Political Geography of the Yucatan Maya. *Carnegie Institution of Washington,* Publ. 613. Washington, D.C.

Ruz Lhuillier, Alberto
1969 La costa de Campeche en los tiempos prehispánicos. *Memorias,* XVIII, Instituto Nacional de Antropología e Historia, México, D. F.

Sabloff, Jeremy A.
1971 (Review of:) *Maya History and Religion,* by J.E.S. Thompson. *American Anthropologist,* Vol. 73, No. 4, pp. 915–17.

Sabloff, Jeremy A. and W. L. Rathje
1975a A Study of Changing Pre-Columbian Commercial Systems. The 1972–1973 Seasons at Cozumel, Mexico. *Monographs of the Peabody Museum,* No. 3. Harvard University, Cambridge, Mass.
1975b The Rise of a Maya Merchant Class. *Scientific American,* Vol. 233, No. 4, pp. 72–82.

Sabloff, Jeremy A. and G. R. Willey
1967 The Collapse of Maya Civilization in the Southern Lowlands: A Consideration of History and Process. *Southwestern Journal of Anthropology,* Vol. 23, No. 4, pp. 311–36.

Sahagun, Fray Bernardino de
1975 *Historia general de las cosas de la Nueva España.* Editorial Porrua, México, D. F.

Sánchez, Jesus B.
1915 *Elementos de historia de Chiapas.* Mexico.

Sanders, William T.
1977 Enviromental Heterogeneity and the Evolution of Lowland Maya Civilization. In *The Origins of Maya Civilization,* edited by R. E. W. Adams, pp. 287–97. University of New Mexico Press, Albuquerque.

Sanders, William T., J. B. Parsons and R. S. Santley
1979 *The Basin of Mexico. Ecological Processes in the Evolution of a Civilization.* Academic Press, New York.

Sapper, Karl
1897 *Das Nordliche Mittel-Amerika nebst einen ausflung nach dem Hochland von Anahuac. Reisen und Studien aus den Jahren 1888–1895.* Druck und Verlag von Friedrich Vieweg und Sohn. Braunschweig.
1901 Die Alta Verapaz (Guatemala). *Mitteilungen der Geographischen Gesselschaft in Hamburg,* Vol. 18, pp. 78–224. Hamburg.

Sarabia Viejo, María Justina
1978 El estanco de la sal en Yucatán (1591–1610). *Anuario de Estudios Americanos,* Vol. 35, pp. 379–405.

Sarg, Franz
1939 Alte Erinncrungen an die Alta Verapaz. In *Deutschtum in der Alta Verapaz: Erinnerungen,* edited by Martin Frey, Compiler, pp. 9–44. Deutche Ausland Institut, Deutschen Verlags Anstalt, Stutgart.

Saul, Frank P.
1972 The Human Skeletal Remains of Altar de Sacrificos. An Ethnobiographic Analysis. *Papers of the Peabody Museum,* Vol. 63, No. 2, Harvard University, Cambridge, Mass.

1973 Disease in the Maya Area: The Pre-Columbian Evidence. In *The Classic Maya Collapse,* edited by T.P. Culbert, pp. 301–24. A School of American Research Book, University of New Mexico Press, Albuquerque.

Scholes, F. V., C. R. Menéndez, J. I. Rubio Mañé and E. B. Adams (Eds.)
1936–1938 *Documentos para la historia de Yucatán.* 3 Vols. Companía Tipográfica Yucateca, Mérida.

Scholes, F. V. and R. L. Roys
1948 The Maya-Chontal Indians of Acalan-Tixchel: A Contribution to the History and Ethnography of the Yucatan Peninsula. *Carnegie Institution of Washington,* Publ. 560, Washington, D.C.

Scholes, F. V. and D. Warren
1965 The Olmec Region at Spanish Contact. *Handbook of Middle American Indians,* Vol. 3, pp. 776–87. University of Texas Press, Austin.

Schott, Arthur
1866 Die Kustenbildung des Nordlichen, Yucatan. *Petermann's Mittheilungen,* Vol. 12, pp. 127–30. Gotha.

Scrimshaw, Nevin S. and Carlos Tejada
1970 Pathology of Living Indians as Seen in Guatemala. *Handbook of Middle American Indians,* Vol. 9, pp. 203–25. University of Texas Press, Austin.

Sharer, Robert J.
1977 The Maya Collapse Revisited: Internal and External Perspectives. In *Social Processes in Maya Prehistory,* edited by N. Hammond, pp. 531–52. Academic Press, New York and London.

n.d. Lower Central America as seen from Mesoamerica. In *The Archeology of Lower Central America,* edited by Fred Lange and Doris Stone, University of New Mexico Press, Albuquerque. [Probable November 1983 publication].

Shattuck, George C.
1933 The Peninsula of Yucatan, Medical, Biological, Meteorological and Sociological Studies. *Carnegie Institution of Washington,* Publ. 431. Washington, D.C.

Shook, Edwin M.
1949 Historia arqueológica de Puerto de San José, Guatemala. *Antropología e Historia de Guatemala,* Vol. 1, No. 2, pp. 3–22. Guatemala City.
1952 Lugares arqueológicos del altiplano meridional central de Guatemala. *Antropología e Historia de Guatemala,* Vol. 4, No. 12, pp. 3–40. Guatemala City.
1965 Archaeological Survey of the Pacific Coast of Guatemala. *Handbook of Middle American Indians,* Vol. 2, pp. 180–94. University of Texas Press, Austin.

Simmons, Michael P. and G. F. Brem
1979 The Analysis and Distribution of Volcanic Ash-Tempered Pottery in the Lowland Maya Area. *American Antiquity,* Vol. 44, No. 1, pp. 79–91.

Sisson, Edward S.
1972 A Report of the Second Session of the "Proyecto Coxcatlan." Salt Production. Paper presented to 71st Annual Meeting of the American Anthropological Association, Toronto, Ontario, December 2, 1972.
1973 First Annual Report of the Coxcatlan Project. *Tehuacan Projects,* Report 3. R. S. Peabody Foundation for Archaeology, Phillips Academy, Andover, Mass.

Siverts, H.
1969 Ethnic Stability and Boundary Dynamics in Southern Mexico. In *Ethnic Groups and Boundaries,* edited by F. Barth, pp. 101–16. Geoge Allen & Unwin, London.

Smith, A. Ledyard
1955 Archaeological Reconnaissance in Central Guatemala. *Carnegie Institution of Washington,* Publ. 608, Washington, D.C.

Smith, A. L. and A. V. Kidder
1951 Excavations at Nebaj, Guatemala. *Carnegie Institution of Washington,* Publ. 594, Washington, D.C.

Solís Preciat, Francisco
1969 *La economía del estado de Yucatán.* Coleción de Estudios Económicos Regionales, Investigación del Sistema Bancos de Comercio. México, D. F.

Stadleman, Raymond
1940 Maize Cultivation in Northwestern Guatemala. *Contributions to American Anthropology and History,* No. 33. *Carnegie Institution of Washington,* Publ. 523, pp. 83–265. Washington, D.C.

Stark, Barbara L.
1974 Geography and Economic Specialization in the Lower Papaloapan, Veracruz, Mexico. *Ethnohistory,* Vol. 21, No. 3, pp. 199–221.

Stoll, Otto
1886 *Guatemala. Reisen und Schilderungen aus den Jahren 1878–1883.* F. A. Brockhaus, Leipzig.

Stone, D. Z.
1966 Synthesis of Lower Central American Ethnohistory. *Handbook of Middle American Indians,* Vol. 4, pp. 209–42. University of Texas Press, Austin.

Stresser-Pean, Guy
1971 Ancient Sources on the Huasteca. *Handbook of Middle American Indians,* Vol. 11, pp. 582–601.

Suárez Molina, Victor M.
1977 *La evolución económica de Yucatán a través del siglo XIX* (2 vols.) Ediciones de la Universidad de Yucatán, México, D. F.

Suasnavar, José
1964 Informe que sobre la erección de un estado compuesto con los pueblos de Los Altos, dió al Gobierno S. de la Nación en 27 de abril de 1824. *Anales de la Sociedad de Geografía e Historia,* Vol. 37, Nos. 1–4, pp. 111–22. Guatemala City.

Termer, Franz
1927 Berichte über Reisen in Mittelamerika. *Mitteilungen der Geographischen Gessellschaft in Hamburg,* Vol. 38. Hamburg.

1931 Zur Archaologie von Guatemala. *Baessler Archiv,* Vol. 14, pp. 167–91.

1941 Zur Geographic der Republik Guatemala II. Beitrage zur kultur-und Wirtschafts-geographic von kittel-und Sudguatemala. *Mitteilungen der Geographischen Gessellschaft in Hamburg,* Vol. 47, pp. 10 – 262. Hamburg.

1957 *Etnología y etnografía de Guatemala.* Editorial del Ministerio de Educación Pública, Guatemala City, pp. 78–79.

Thompson, J. E. S.
1953 Relaciones entre Veracruz y la Región Maya. In *Huastecos, Totonacos y sus Vecinos,* edited by I. Bernal y E. Dávalos Hurtado, *Revista Mexicana de Estudios Antropológicos,* Vol. 13, Nos. 2–3, pp. 447–54. México, D. F.

1964 Trade Relations Between the Maya Highlands and Lowlands. *Estudios de Cultura Maya,* Vol. 4, pp. 13–49. México, D. F.

1965 Archaeological Synthesis of the Southern Maya Lowlands, *Handbook of Middle American Indians,* Vol. 2, pp. 331–59. University of Texas Press, Austin.

1966 *The Rise and Fall of Maya Civilization.* University of Oklahoma Press, Norman.

1970 *Maya History and Religion.* University of Oklahoma Press, Norman.

Tourtellot, Gair and J. A. Sabloff
1972 Exchange Systems Among the Ancient Maya. *American Antiquity,* Vol. 37, No. 1, pp. 126–35.

Tovilla, Capt. Martin Alfonso
1960 *Relación histórica descriptiva de las provincias de la Verapaz y de la del Manché, escrita por el Capitán Don Martín Alfonso Tovilla. Año de 1635.* Edited with introduction and paleogra-

phy by F. V. Scholes and E. B. Adams. Editorial Universitaria, Guatemala City.

Tozzer, A.M.
1941 Landa's relación de las cosas de Yucatán: A Translation. *Papers of the Peabody Museum,* No. 18. Harvard University, Cambridge, Mass.

Triay Peniche, Manuel
1977 La industria salinera de Yucatán. *Diario de Yucatán,* February 19, 1977. Mérida.

Urbina, Fernando
1909 Algunas observaciones acerca de la geografía física del Estado de Yucatán. *Boletín de la Sociedad Geológica Mexicana,* Vol. 5, pp. 91–101. México, D. F.

Valencia Bellavista, Juan M.
1978 Maxcanú. Estudio socio-económico de un centro de mayor captacion social en áreas rurales vista a través de la teoría de los centros rectores. Tesis Profesional. Escuela de Ciencias Antropológicas de la Universidad de Yucatán, Mérida.

Van Loon, H. W.
1937 The Story of Salt. *Natural History,* Vol. 39, pp. 78–98.

Valásquez Valadéz, Ricardo
1980 Recent Discoveries in the Caves of Loltun, Yucatan, Mexico. *Mexicon,* Vol. 2, No. 4, pp. 53–55.

Velson, Joseph S. And Thomas C. Cook
1975 Transport of Stone Monuments to the La Venta and San Lorenzo Sites. *Contributions of the University of California Archaeological Research Facility,* No. 24, pp. 1–39. Berkeley.

Viana, F. de, L. Gallego y G. Cadena
1955 Relación de la provincia de la Verapaz, hecho por los religiosos de Santo Domingo de Cobán (1574). *Anales de la Sociedad de Geografía e Historia de Guatemala,* Vol. 28, Nos. 1–4, pp. 18–31. Guatemala City.

Villacorta, Rafael
1883 *El tercer informe del comisionado J. Carlos Manó y las salinas de los Nueve Cerros.* Tipografía de "El Quetzal," Cobán.

Villa Rojas, Alfonso
1945 The Maya of East Central Quintana Roo. *Carnegie Institution of Washington,* Publ. 559. Washington, D.C.

1946 Notas sobre la etnografía de los Indios Tzeltales de Oxchuc, Chiapas, México. *Microfilm Collection of Manuscripts on Middle American Cultural Anthropology,* No. 7. University of Chicago, Chicago.

Vlcck, David T., Silvia Garza T. de González and E. B. Kurjack
1978 Contemporary Farming and Ancient Maya Settlements: Some Disconcerting Evidence. In *Prehispanic Maya Agriculture,* edited by P. D. Harrison and B. L. Turner, pp. 211–23. University of New Mexico Press, Albuquerque.

Vogt, Evon Z.
1969 *Zinacantan. A Maya Community in the Highlands of Chiapas.* Harvard University Press, Cambridge, Mass.

Walters, Gary Rex (Ed.)
1977 Preliminary Results of the Unul Project. *Museum of Anthropology Special Reports,* No. 1. University of Missouri-Columbia, Columbia.

Warren, Bruce W.
1978 The Sociocultural Development of the Central Depression of Chiapas, Mexico: Preliminary Considerations. Doctoral Dissertation, University of Arizona, Tucson.

Wasserstrom, Robert F.
1970 Our Lady of the Salt. Bachelor's Honors Thesis, Committee on Folklore and Mythology, Harvard University, Cambridge, Mass.

1976 La evolución de la economía regional en Chiapas: 1528–1975. *América Indigena,* Vol. 36, No. 3, pp. 479–98.

Weaver, Muriel Porter
1972 *The Aztecs, Maya, and their Predecessors Archaeology of Mesoamerica.* Seminar Press, New York and London.

Webb, Malcolm C.
1973 The Peten Maya Decline Viewed in the Perspective of State Formation. In *The Classic Maya Collapse,* edited by T. P. Culbert, pp. 367–404. A School of American Research Book, University of New Mexico Press, Albuquerque.

1975 The Flag Follows Trade: An Essay on the Necessary Interaction of Military and Commercial Factors in State Formation. In *Ancient Civilizations and Trade,* edited by J. A. Sabloff and C. C. Lamberg-Karlovsky, pp. 155–209. University of New Mexico Press, Albuquerque.

Webster, David L.
1977 Warfare and the Evolution of Maya Civililzation. In *The Origins of Maya Civilization,* edited by R. E. W. Adams, pp. 335–72. University of New Mexico Press, Albuquerque.

1979 Cucá, Chacchob, Dzonot Aké—Three Walled Northern Maya Centers. *Occasional Papers in Anthropology,* No. 11. Department of Anthropology, Pennsylvania State University, University Park.

Whittlesey, Marietta
1978 *Killer Salt.* Avon Books, New York.

Willey, G. R., W. R. Bullard, Jr., J. B. Glass and J. C. Gifford
1965 Prehistoric Maya Settlements in the Belize Valley (2 vols.) *Papers of the Peabody Museum,* Vol. 54. Harvard University, Cambridge, Mass.

Willey, G. R., T. P. Culbert and R. E. W. Adams
1967 Maya Lowlands Ceramics: A Report from the 1965 Guatemala City Conference. *American Antiquity,* Vol. 32, pp. 289–315.

Williams, J. J.
1852 *The Isthmus of Tehuantepec.* D. Appleton & Co., New York.

Wilson, Eugene M.
1980 Physical Geography of the Yucatan Peninsula. In *Yucatan. A World Apart,* edited by E. H. Moseley and E. D. Terry, pp. 5–40. University of Alabama Press, Tuscaloosa.

Wing, Elizabeth S. and David Steadman
1980 Vertebrate Faunal Remains from Dzibilchaltun. In *Excavations at Dzibilchaltun, Yucatan, Mexico,* by E. W. Andrews IV and E. W. Andrews V. *Middle American Research Institute,* Publ. 48, pp. 326–31. Tulane University, New Orleans.

Ximénez, Fray Francisco
1929–1931 Historia de la Provincia de San Vicente de Chiapa y Guatemala. (Orig. 1720). 3 vols. Biblioteca "Goathemala," Vols. 1–3, *Sociedad de Geografía e Historia,* Tipografía Nacional, Guatemala City.

1965 Historia de la provincia de San Vicente de Chiapa y Guatemala (2nda edición). (4 Vols.) *Biblioteca Guatemalteca de Cultura Popular,* Vols. 81–84, Editorial "José de Pineda Ibarra." Guatemala City.

Zabala Cubillos, Manuel
1959 Field Notes on Ixtapa, Chiapas. Ms. on file in Harvard Chiapas Project, Harvard University, Cambridge, Mass.

Zaráte, José C.
1917 Las salinas de México y la industria de la sal común. *Anales del Instituto Geológico de México,* Vol. 1, No. 2. México, D. F.

Index

Acajutla Region, El Salvador, 104–5
Acalapa, Veracruz, 47, 120
Acapán, Guatemala. *See* Salinas Acapán
Acapetagua, Chiapas, 63, 65
Aguada Grande, Cozumel, 45
Ah Canul Province, Yucatán, 34, 136, 137
Ah Kin Chel Province, Yucatán, 136, 137
Alta Verapaz, Guatemala (Department), 7, 18, 76, 77, 95, 100, 101. *See also* Salinas de los Nueve Cerros
Altamira, Chiapas, 66
Altamira, Tamaulipas ("Lomas del Real;" "Salinas San Enrique"), 50
Altar de Sacrificios, Guatemala, 100, 122, 127
Alvarez, Chiapas, 66
Amatitlán, Guatemala, 15, 77–79, 112, 120, 125
Ambergris Cay, Belize, 18, 42, 46–47, 120, 138
Andrews IV, E. Wyllys, 8, 123, 124, 129
Apastes. *See* Ceramics
Arriaga, Chiapas, 62, 63–65
Asanyamba, El Salvador, 107
Asoleaderos (solar evaporation tanks), 70–71, 74, 102, 104–5, 109
Atzam, Chiapas. *See* Salinas Atzam
Atzam, Guatemala. *See* San Miguel Acatán
Ayacachapa, El Salvador. *See* Salinas de Ayacachapa
Aztecs: salt deity, 113; salt glyph, 3; salt made from urine, 11, 21; salt used in healing, 12; saltmakers' guilds, 113

Baja Verapaz, Guatemala (Department), 76, 77, 93, 101. *See also* Cobán
Ball, Joseph W., 8, 30, 31, 32, 122, 123, 124, 128
Barillas, Guatemala, 87, 88–89, 93, 111, 129
Barrios, Rufino, 94–95
Becán, Campeche, 118, 122, 133
Belize: Itzá presence in, 129; salinas of, *see* Ambergris Cay; salt imports, from Yucatán, 34, 37, 38, 39, 41, 123, 130, 135; from Britain, 38, 101, 130; from El Salvador, 107; source of British salt in Guatemala and Quintana Roo, 38, 75, 77, 101
Benequenes (salt loaves), 58, 60
Berlín, Chiapas, 64
Black salt of Sacapulas, 89, 92
Blom, Frans, 4, 5, 53, 54, 55
Bolas (salt balls), 89
Bolontehuitz, Guatemala. *See* Salinas de los Nueve Cerros
Brine springs. *See* Salt brine springs
British salt: in Belize, 101, 130; in Guatemala, 75, 77, 101, 115–16, 130; in Quintana Roo, 38, 77, 101

Cabeza de Toro, Chiapas, 64
Cacaxtla, Tlaxcala, 127
Cajones. *See Coladeras*
Cakchiquel Indians, Guatemala, 93, 118, 129
Calvario, Guatemala, 84, 85
Campeche: city of, 35, 37; salinas of, 26, 42; salt trade, 39, 49, 50, 117
Campo de Aviación, Chiapas, 59
Canbalám, Yucatán, 33
Canoes, used for filtering brine. *See Coladeras*
Capulín, Chiapas, 64
Cargo system, role of salt in, 11
Casas Viejas, Guatemala, 76
Catepan, Guatemala, 85
Caucel, Yucatán. *See* Xtul
Cehache Indians, Campeche-Guatemala, 41

Cehpech Province, Yucatán, 137
Celestún, Yucatán, 26, 31, 33, 36, 37, 38, 41, 124, 137
Ceramics: clay cylinders supports for brine-cooking vessels, 90, 111; containers for cooking salt brine, 16, 59, 61, 68–69, 77, 79, 82, 84, 88, 89, 90, 99, 102, 104, 111; containers for transporting and storing salt brine, 59, 61, 82, 83, 84, 97, 99; diagnostic chronological types, 62, 123, 127; tradewares, 32, 123, 126, 127, 132
Cerro Chiblac, Guatemala, 88
Cerros, Belize, 122, 123, 133, 135
Chakán Province, Yucatán, 137
Champerico, Guatemala, 69, 71, 73, 74, 76, 77, 122
Champotón, Campeche, 42
Chamula, Chiapas, 61
Chapán, Guatemala. *See* Salinas Chapán
Chauaca (or Chikinchel) Province, Yucatán, 137
Chavihau (Salinas de Yobaín; Yobaín Beds), Yucatán, 28, 38, 137
Chiapa de Corzo, Chiapas, 61
Chiapas: salinas of highlands, 51–62, 110–18, 120–21, 122, 129, 131; salinas of Pacific coast, 15, 62–67, 110–17, 120, 122; salt imports, 41, 49, 60, 64, 67, 87, 100; salt production, 5, 56, 60–61, 63–67, 115; salt trade, 56, 60, 61, 67, 115
Chiapanecs, 59–60, 129, 135, 137
Chichen Itzá, Yucatán, 8, 32–33, 128–29
Chicxulub, Yucatán, 28, 38, 137
Chila, Veracruz, 50
Chilapchi, Guatemala. *See* Chixjoj
Chilón, Guatemala, 84
Chikinchel Province, Yucatán. *See* Chauaca
Chimaltenango, Guatemala (Department), 76, 79
Chinkultic, Chiapas, 85
Chiquimula, Guatemala (Department), 76, 77
Chiquimulilla, coast of. *See* Guazacapán, coast of
Chixjoj (or Chexjoj, or Chilapchi, or Santa Delia), Guatemala, 82, 87
Chontal Maya, 127
Chuburná, Yucatán, 28, 38, 137
Chuj Indians, Guatemala. *See* San Mateo Ixtatán
Chumhabín, Yucatán, 32
Chunchucmil, Yucatán, 124
Chutinamit, Guatemala, 91
Civit Krus, Chiapas, 62
Coatzalcoalcos, Veracruz, 16, 42, 49, 60, 64, 67
Cobá, Quintana Roo, 122, 126, 128–29
Cobán, Guatemala, 93, 98, 99, 101
Coladeras (or *cajones,* used for filtering brine): at Amatitlán, Guatemala, 77; at Sacapulas, Guatemala, 89, 90; use of canoes on Pacific Coast of Guatemala and El Salvador, 16, 68, 74, 102. *See also* Tapesco method
Comitán, Chiapas, 56, 113, 118
Communal ownership of salinas. *See* Tenure of salinas
Containers for transporting salt and brine, 3, 23, 44, 59, 82, 83. *See also* Ceramics
Contreras Alto, Guatemala, 78, 79
Costa Rica: clay cylinders, 111; salinas of, 6, 111; salt imports from El Salvador, 107
Cozumel Island. *See* Isla Cozumel
Crosses, found at highland salinas, 11, 56, 62, 85–86
Cuba, salt imports, 3, 35, 37, 38, 39, 130
Cuymatlan, Veracruz, 50
Custepeques, Chiapas. *See* La Concordia
Cylinders, supports for brine-cooking vessels, 90, 111

De la Peña, Moisés, 5, 51, 56, 60, 63, 64, 66, 67, 138
Diana Milán, Yucatán, 28
Diet, role of salt in, 1, 8–10, 122, 123. *See also* Salt-free diet
Dillon, Brian D., 5, 7, 95–97, 99, 100, 101, 134
Dolores, Yucatán, 28
Dzemul, Yucatán. *See* Xtampú
Dzibilchaltún, Yucatán, 8, 10, 124, 133
Dzidzantún, Yucatán. *See* Santa Clara and Mina de Oro
Dzilam de Bravo, Yucatán, 30
Dzilam González, Yucatán, 124

Eaton, Jack D., 7, 23, 30, 31, 32, 109
Ecab, Quintana Roo, 34, 44, 137
Edzná, Campeche, 122
El Chayal, Guatemala, 126
El Cuyo, Yucatán, 28, 30, 31, 32, 33, 38, 42
El Pajón, Chiapas, 7, 66

El Porvenir, Guatemala. *See* Nucá Atzam; *see also* San Mateo Ixtatán
El Progreso, Guatemala. *See* Nucá Atzam
El Quetzal, Guatemala. *See* Nucá Atzam
El Salitre, Guatemala. *See* Amatitlán
El Salvador: salinas of, 102–7, 108, 109, 110, 121; salt exports, 16, 75, 107, 115–16; salt production of, 39, 107, 115
El Tamarindo, El Salvador, 107
El Tihuilote, El Salvador, 104
Emal, Yucatán, 11, 28, 31, 32, 33, 34, 124. *See also* Las Coloradas
Escuintla, Chiapas, 63, 65, 66
Escuintla, Guatemala (Department), 76

Fanega, measure of capacity, 39, 40, 42, 45, 136–37
Feldman, Lawrence H., 7, 18, 59, 60, 76, 78, 79, 101

Garita Chapina, Guatemala, 73, 76
Golfo Dulce, Guatemala, 77
Guamuchil (or Huamuchil), Chiapas, 66
Guatemala: salinas of the highlands, 77–95; salinas of Pacific coast, 15, 68–77; salinas of the lowlands, *see* Salinas de los Nueve Cerros; salt imports from Belize (British salt), 75, 77; salt imports from El Salvador, 16, 75, 76, 107, 115–16; salt imports from other countries, 75; salt production, 5, 39, 75–76, 79, 86–89, 92, 94–95, 99–100, 102, 107, 108–31
Guatemala (Department), 76
Guayatengo (or Huayalotengo), Chiapas, 66
Guazacapán, coast of (or coast of Chiquimulilla), 76
Gulf Coast Region, Mexico: salinas of, 47–50; Yucatecan salt imports, 35–42, 47–50, 130. *See also* Tabasco, Veracruz, Tamaulipas
Gulf of Honduras, 18, 41

Hacienda San Pablo, Chiapas, 65
Halite (rock salt) deposits, 49, 88
Holbach, Yucatán, 33
Holbox Island. *See* Isla Holbox
Honduras, 6, 10, 15, 18, 34, 35, 36, 41, 107, 116
Hornos (iron vat ovens), 16, 58, 59, 70, 71, 74, 75, 88, 94, 95, 99, 102, 105, 111
"Hot-Cold" Syndrome, 13
Huamuchil, Chiapas. *See* Guamuchil
Huayalotengo, Chiapas. *See* Guayatengo
Huehuetenango, Guatemala, 86, 87, 92, 93, 94, 118. *See also* Zaculeu
Huehuetenengo, Guatemala (Department), 76, 77, 81, 87
Huixtla, Chiapas, 63, 65
Huixtocihuatl, Aztec salt deity, 113
Hul, Guatemala. *See* San Mateo Ixtatán

Icaiché Indians, Quintana Roo, 38
Industria Salinera de Yucatán, S.A., 29, 38, 115
Iron cauldrons. *See Peroles*
Iron vat ovens. *See Hornos*
Isla Cerritos, Yucatán, 33
Isla Cozumel, Quintana Roo, 42, 44, 45–46, 120, 126
Isla del Cármen, Campeche, 42, 120
Isla Holbox, Quintana Roo, 42, 43, 44, 138
Isla Mujeres, Quintana Roo, 34, 42, 44–45, 137
Itzá: in northern Yucatán, 8, 32–33, 128–29, 135; Petén Itza, 98, 100
Ixil Indians, Guatemala, 93, 94
Ixtán, Guatemala. *See* Salinas de Ixtán
Ixtapa, Chiapas, 14, 51, 56–61, 111, 112, 113, 120, 129, 131
Ixtepeque, Guatemala, 126
Izabal, Guatemala (Department), 77, 101
Iztapa, Guatemala, 72, 76, 125
Iztapa, Tabasco, 47

Jalapa, Guatemala (Department), 76
Jaltepeque Region, El Salvador, 105
Jiquilisco Region, El Salvador, 105–6
Jolom Atzam (Yulconob), Guatemala, 88
Jones, Christopher, 125, 127, 133, 134
Jutiapa, Guatemala (Department), 76

Kaminaljuyú, Guatemala, 125
Kanjobal Indians, Guatemala. *See* San Miguel Acatán
Komchén, Yucatán, 124, 133

La Concordia (Custepeques), Chiapas, 14, 56, 112, 113, 116, 118

La Gloria, Chiapas, 64
La Providencia, Guatemala, 88
La Providencia, Yucatán, 28, 31, 32
La Soledad, Guatemala. *See* La Providencia
La Unión Region, El Salvador, 106–7
Lacandón Indians, Chiapas, 20, 98, 100
Laguna de Términos, Campeche, 42
Laguna La Joya, Chiapas, 64
Laguna Redonda, Chiapas, 66
Lake Amatitlán, Guatemala. *See* Amatitlán
Lake Izabal, Guatemala, 101, 115
Lamanai, Belize, 122, 133, 135
Landa, Diego de, 3, 10, 12, 14, 33, 34–35
Las Coloradas, Yucatán, 23, 28–29, 30, 38, 109. *See also* Emal
Las Victorias, Guatemala, 93, 95
Llano de las Animas, Guatemala, 78–79
Lomas del Real, Tamaulipas. *See* Altamira
Los Jicaques, Guatemala, 78, 79
Los Nogales, Guatemala. *See* La Providencia
Lozano García, Raúl, 4, 39, 42, 44–50, 55, 64, 66

McBryde, Felix W., 4, 5, 10, 14, 70, 71, 87, 92
McVicker, Donald E., 59, 61
Magaña, Ausencio, 45
Mam Kingdom, Guatemala, 85, 86, 129. *See also* Zaculeu
Mapastepec, Chiapas, 62, 63, 65, 66
Mar Muerto Lagoon, Chiapas, 64
Matamoros, Tamaulipas, 50
Mayapán, Yucatán, 11, 33, 34, 35, 129
Mazatán, Chiapas, 63, 65, 66
Mendizábal, Miguel Othón de, 4, 5, 8, 14, 42, 44, 47, 49, 50, 55, 60, 62, 64, 66, 108, 135
Merchants. *See* Salt merchants, Trade
Mexico: central Mexican salt imports from Yucatán, 3, 15, 35, 37, 38, 40, 124, 130; central Mexican salt industries, 4, 15, 110, 113, 124. *See also* Aztecs, Cacaxtla, Oaxaca, Salt uses, Tabasco, Tamaulipas, Tehuacán, Tenochtitlán, Teotihuacán, Veracruz, Xochicalco
Mina de Oro (Salinas de Dzidzantún), Yucatán, 27, 28, 38, 137
Mirador, Guatemala, 122
Moctezuma, Chiapas, 64
Monterrico, Guatemala, 73, 76
Montes, Felix, 72, 74, 75

Nebaj, Guatemala, 92, 93, 94
Nicaragua, salt imports from El Salvador, 107
Nucá Atzam, Guatemala, 88
Nueve Cerros, Guatemala. *See* Salinas de los Nueve Cerros
Nutrition and salt. *See* Diet

Oaxaca: prehispanic salt consumption, 10; salt sources in, 7, 16, 120
Ocós, Guatemala, 6, 69, 71, 73, 76, 90, 111, 122

Pajaritos, Veracruz, 49
Paillés, Maricruz, 7, 66, 120
Palms, as source of salt. *See* Salt sources
Panama, salt imports from El Salvador, 107
Pangas (iron vats). *See Hornos*
Pánuco, Veracruz, 35, 36, 41, 49–50
Paredón, Chiapas, 64, 66
Pasaco, Guatemala: coast of, 76
Paso del Cerro, Yucatán, 30, 33
Patios (solar evaporation pans). *See* Solar evaporation process
Peroles (iron cauldrons), 16, 69–70, 74, 102, 104, 111
Petén, Guatemala (Department), 3, 10, 20, 41, 76, 77, 97–101, 118, 122, 126, 130
Petén Itza. *See* Itzá
Pilas (solar evaporation pans). *See* Solar evaporation process
Pilones (salt cakes), 82, 84
Pink salt "of Comitán" (from La Concordia), 54, 56
Pirate activities and salt taxation, 37, 74
Playa Jiménez, Tabasco, 47
Playa Lavaderos, Guatemala. *See* Amatitlán
Ponce, Flavio, 99, 100
Population: prehispanic Maya highlands, 115; prehispanic Maya lowlands, 10, 114; rising density during Formative period, 19, 21, 122
Portatenco, Chiapas, 53, 54, 55, 120
Progreso, Yucatán, 28, 41
Property rights to salinas. *See* Tenure of salinas
Puerto Arista, Chiapas, 64
Puerto Madero, Chiapas, 62, 63, 65, 66

Punta Boxcohuo, Yucatán, 26, 28, 137
Punta Cerrito, Yucatán, 28, 30, 33
Punta Choventún, Yucatán, 27, 28, 137
Punta de Agua, Chiapas, 66
Punta Desconocida (Punta Nimún), Yucatán, 26
Punta Flor, Chiapas, 64–65
Punta Palmar, Yucatán, 30
Putún, 127
Puuc Region, Yucatán-Campeche, 127

Quesitos (salt cakes), 40
Quetzaltenango, Guatemala (Department), 76, 87
Quiché, Guatemala (Department), 76, 77, 89–95
Quiché Indians, Guatemala, 89, 91
Quiché Kingdom, Guatemala, 91, 93, 129, 135
Quiñonez, José, 5, 79, 82, 84, 88
Quiahuistlán, Veracruz, 49
Quintana Roo: imports of British salt, 38, 77, 101; salinas of, 42–46, 137, 138

Rathje, William L., 8, 47, 98, 125, 126, 127, 133, 134, 135
Real de Salinas, Yucatán, 40
Retalhuleu, Guatemala (Department), 76
Río Lagartos, Yucatán, 28, 29, 30, 31, 33, 35
Rivers, mentioned, Belize, 41; Candelaria, 41; Chixoy, 7, 95; Dulce, 41; Grijalva, 41; Hondo, 41; Lacantún, 100; Motagua, 41, 126; New, 41, 123; Pánuco, 41; Pasión, 95, 100, 116; Negro, 89, 91, 92; Salinas, 95, 100; San Pedro Mártir, 41; Sarstoon, 41; Subín, 100; Usumacinta, 7, 41, 91, 95, 100, 116, 117, 118, 127, 135; used in salt trade, 113, 118, 135
Roche family, Yucatán, 38, 39, 44
Roys, Ralph L., 4, 5, 34, 41, 42, 44, 46, 98, 113, 118–19, 124, 130, 136

Sabalpop, Guatemala, 79
Sacapulas, Guatemala, 4, 11, 12, 14, 15, 81, 89–93, 94, 95, 101, 111, 112, 115, 118, 121, 129, 131, 135
Sacatepequez, Guatemala (Department), 76
Saharan salt trade, 119, 126
Sal cocida process: 16–18, 110–13; antiquity of, 6, 66, 108, 110–11; in Chiapas, 56–67; in El Salvador, 102–107; in Guatemala, 68–95; in Veracruz, 49
Sal solar. See Solar evaporation process
Salinas Acapán, Guatemala, 69, 71
Salinas Atzám, Chiapas, 11, 51, 56, 61–62, 111, 112, 131
Salinas Chapán, Guatemala, 71
Salinas de Ayacachapa, El Salvador, 104–5
Salinas de Caucel, Yucatán. *See* Xtul
Salinas de Dzidzantún, Yucatán. *See* Santa Clara and Mina de Oro
Salinas de Dzemul, Yucatán. *See* Xtampú
Salinas de Ixtán, Guatemala, 71
Salinas de Magdalena, Guatemala, 70, 93–95, 111
Salinas de los Nueve Cerros, Guatemala, 7, 18, 77, 95–101, 111, 112, 115, 116, 118, 121, 127
Salinas de Portatenco, Chiapas. *See* Portatenco
Salinas de Punta Flor, Chiapas. *See* Punta Flor
Salinas de Santa Marta, Guatemala, 72
Salinas de Santa Rosa, Guatemala, 72
Salinas de Sinanché, Yucatán. *See* San Crisanto
Salinas de Yobaín, Yucatán. *See* Chavihau
Salinas El Aceituno, El Salvador, 106
Salinas El Porvenir, Guatemala. *See* Nucá Atzam
Salinas San Enrique, Tamaulipas. *See* Altamira
Salinitas, El Salvador, 104
Salt-free diet, 1
Salt brine springs, 18, 111; in Chiapas, 51–62, 111; in Guatemala, 79–100, 111; in Veracruz, 47–48
Salt domes. *See* Coatzalcoalcos, Halite, Pajaritos
Salt consumption in diet, 1, 8–10, 122, 123
Salt merchants, 3, 60, 61, 92, 100, 113, 119, 125, 126, 127, 130, 131, 133; Gulf Coast merchants, 32–33, 127–28; Itzá traders, 128–29; Yucatecan nobles as traders, 113, 119, 129
Salt playas. *See* Acalapa, Amatitlán, Quiahuistlán, Sacapulas, Salinas de los Nueve Cerros
Salt production methods. *See* Sal cocida process, Seasonality, Solar evaporation process

Salt production statistics: Belize, 46; Chiapas, 57, 60, 61, 67, 114–15; El Salvador, 5, 39, 107, 109, 115; Guatemala, 5, 39, 75–76, 77, 79, 86–87, 88, 92, 94–95, 99–100, 109, 115; Gulf Coast region, 47–50; Maya area, 10, 114–15; Yucatán, 29, 38, 39–41, 42, 44–45, 115, 136–38
Salt sources (other than salinas): animal meat, 18–19, 21, 22; plant sources, 19–21, 130; sea water, 18, 21; urine, 11, 21
Salt storage, 25, 35, 97, 99, 112
Salt trade: antiquity and origins of, 31, 118, 122–23; studies of, 3–8; trading spheres, 116–18. *See also* Belize, British salt, Chiapas, El Salvador, Guatemala, Gulf Coast, Rivers, Salt merchants, Transportation of salt, Yucatán
Salt uses: artifacts made of, 14, 53–54; currency, 3, 13–14; industrial, 14–16; livestock consumption, 15, 64, 75, 79, 82, 87, 88, 92, 95; medicinal, 3, 11–13, 56, 60–62, 87, 89; mining, silver, 3, 14–15, 37, 38, 130; preservative, 10–11, 42, 44, 49, 54; ritual, 3, 11–13, 53–54, 56, 61–62; tanning, 11, 15; warfare, in protective shields and jackets, 14. *See also* Diet
San Bruno, Yucatán, 28
San Crisanto (Salinas de Sinanché), Yucatán, 25, 28, 31, 32, 34, 38, 137
San Cristóbal de las Casas, Chiapas, 56, 61, 67
San Enrique, Tamaulipas. *See* Altamira
San Fernando, Tamaulipas, 50
San Fernando, Yucatán, 124
San José, Guatemala, 71–74, 76, 77, 122, 125
San Marcos, Guatemala (Department), 76
San Mateo Ixtatán, Guatemala, 11, 56, 81–87, 89, 111, 112, 113, 115, 121, 129, 131
San Miguel, Yucatán, 28
San Miguel Acatán, Guatemala, 79–81, 87, 129
San Miguel Uspantán, Guatemala, 92, 95
Santa Clara (Salinas de Dzidzantún), Yucatán, 28, 38, 137
Santa Cruz del Quiché, Guatemala. *See* Quiché, Guatemala (Department)
Santa Cruz Indians, Quintana Roo, 38
Santa Delia, Guatemala. *See* Chixjoj
Santa Josefa, Guatemala. *See* San Mateo Ixtatán
Santa María de Magdalena, Guatemala. *See* Salinas de Magdalena
Santa Marta, Guatemala. *See* Salinas de Santa Marta
Santa Rosa, Guatemala (Department), 76. *See also* Salinas de Santa Rosa
Santo exchange, role of salt in, 11
Sea water (as source of salt), 18, 21
Seasonality of salt production and trade, 23–25, 53, 63, 70, 89, 112–13
Seibal, Guatemala, 122, 127
Shell celts, used for scraping up salt, 23, 109
Silver mining: role of salt in. *See* Salt uses
Sinanché, Yucatán. *See* San Crisanto
Sipacate, Guatemala, 71, 72, 76
Soconusco Coast, Chiapas, 63, 65–66, 67
Solar evaporation process: 16–18, 108–10; antiquity of, 7, 30–31, 108–9; in Belize, 18, 46; Chiapas, 51–53, 63, 64, 65, 110; El Salvador, 102, 109; Guatemala, 70–71, 99, 109, 112; Gulf Coast Region, 49–50; Yucatán, 22–30, 34–35, 42–45, 108–9
Solar evaporation tanks. *See* *Asoleaderos,* Solar evaporation process
Sololá, Guatemala (Department), 14, 76
Sodium: role in human metabolism, 1, 8–10, 122, 123
Soto la Marina, Tamaulipas, 50
"Stars of David" (salt artifacts), 53, 54, 55
Suchiate River, 62, 65, 66, 68
Suchitepequez, Guatemala (Department), 76
Sumu Indians, Honduras, 21
Surveys of salinas, Colonial and Modern periods, 4–5, 7, 36, 39, 44, 136–38

Tabasco: salinas of, 47; Yucatecan salt imports, 41, 42, 50, 130
Tahuexco, Guatemala, 4, 71, 72, 76
Tamaulipas: salinas of, 49–50; Yucatecan salt imports, 41, 42, 50, 130
Tampico, Tamaulipas, 49–50
Tanchoy, Veracruz, 50

Tapesco method of salt production, 63, 64, 65, 110
Taxes and tributes on salt production, 3; in Chiapas, 62; in Guatemala, 73, 74, 76, 78, 86, 93; in Yucatán, 34, 35, 36–37, 39, 40, 42, 45–46, 136
Tazas (for brine storage), 63
Tehuacán Valley, Puebla, 110, 111
Tehuantepec, salt of, 16, 60, 62, 64, 67
Tenam, Chiapas, 85
Tenure of salinas: communal ownership, 3, 34, 58; conflicts over, 3, 73–74, 91–92; early Spanish takeovers, 3, 34, 36–37, 73–74, 130; in Chiapas, 58; in Guatemala, 73–74, 82, 92, 98; in Yucatán, 34, 35, 36–37, 38, 124, 129; Spanish legislation on, 36–37, 73–74
Teotihuacán, Mexico, 134
Thompson, J. Eric S., 5, 12, 14, 20, 41, 44, 46, 47, 50, 95, 97, 98, 101, 115, 118, 123, 127, 130
Tinajas (water jars). *See* Ceramics
Tikal, Guatemala, 10, 122, 125–26, 132, 133, 134
Tenochtitlán, Mexico (Aztec capital), 11
Tlacotalpan, Veracruz, 49, 50
Toltecs, 127
Tonalá, Chiapas, 62, 63, 64, 66, 67
Toquegua Indians, Guatemala, 18
Totonicapán, Guatemala (Department), 76, 87
Trade: early trade, 31, 118, 122–23, 132; modes of exchange, 131–33; of jade, 126, 129, 132, 134; of obsidian, 31, 123, 126, 132, 134, 135; of other commodities, 14, 61, 116, 124, 126, 129, 130, 132, 134, 135; trade and the rise of states, 133–35. *See also* Ceramics, Saharan trade, Salt merchants, Salt trade, Transportation of salt
Traders. *See* Salt merchants
Transportation of salt, 40, 109, 114, 116; by land, 23, 29, 35, 41, 100, 101, 109, 114, 116, 124; by water, 35, 41, 100, 109, 113, 114, 116, 118, 123, 133, 135. *See also* Containers, Rivers, Salt trade
Tribute of salt. *See* Taxes
Tulum (Zama), Quintana Roo, 46
Tuxtla Gutiérrez, Chiapas, 56, 67
Tzeme, Yucatán, 124
Tzotzil Indians, Chiapas, 56, 60, 61

Uaxactún, Guatemala, 122, 133
Uaymitún, Yucatán, 28
Urbanism, role of salt in the rise of, 8, 124, 125, 133
Uspantán, Guatemala. *See* San Miguel Uspantán

Veracruz: salinas of, 47–50; salt imports, 35, 36, 37, 39, 41, 47, 50, 129, 130
Virgen del Rosario, Salinas Atzám, Chiapas, 11, 62

War, over salt sources, 1, 3, 32, 59–60, 91, 98, 128, 129, 135

Xcambó, Yucatán, 32, 33
Xcaret (Polé), Quintana Roo, 46
Xkalak, Quintana Roo, 46
Xochicalco, Morelos, 127
Xtampú, Yucatán, 11, 28, 32, 33, 34, 38
Xtul (Salinas de Caucel; Caucel Beds), Yucatán, 28, 35, 137
Xul, Guatemala. *See* San Mateo Ixtatán

Yaxchilán, Chiapas, 100
Yobaín Beds, Yucatán. *See* Chavihau
Yola, Guatemala. *See* Nucá Atzam
Yucatán: Caste War, 37, 40; colonial salt trade, 3, 34–37, 39; history of salt production, 34–45, 120, 121; internal salt trade, 3, 31, 35, 41, 117, 124, 130; modern salt industry, 15, 23–30, 37–41; north coast geology, 22–23; prehispanic salt trade, 3, 31–34, 41, 118–35; salinas of, 22–46; salt exports, to Belize, 21, 37, 38, 39, 41, 123, 130, 135; to Cuba, 3, 35, 37, 38, 39, 130; to Guatemala highlands, 115, 116; to Gulf Coast region, 35–42, 47–50, 123, 129, 130; to Honduras, 34, 35, 41, 129; to Petén, Guatemala, 10, 41, 118, 126, 130; salt production statistics, 39–41, 42, 44–45, 115, 136–38; Salt Survey of 1605, 36, 44, 136–38
Yulconob, Guatemala. *See* Jolom Atzam

Zaculeu, Guatemala, 85, 86
Zaráte, José, 4, 15, 39, 40, 42, 44, 47, 49, 50, 55, 60, 62, 64, 66, 138
Zinacantán, Chiapas, 12, 59–62
Zinacantecos, 11, 12, 13, 59, 61, 113, 129, 135